AF594186

Springer Tracts in Advanced Robotics

Volume 43

Pierre Lamon

3D-Position Tracking and Control for All-Terrain Robots

Springer

Professor Bruno Siciliano, Dipartimento di Informatica e Sistemistica, Università di Napoli Federico II, Via Claudio 21, 80125 Napoli, Italy, E-mail: siciliano@unina.it

Professor Oussama Khatib, Robotics Laboratory, Department of Computer Science, Stanford University, Stanford, CA 94305-9010, USA, E-mail: khatib@cs.stanford.edu

Professor Frans Groen, Department of Computer Science, Universiteit van Amsterdam, Kruislaan 403, 1098 SJ Amsterdam, The Netherlands, E-mail: groen@science.uva.nl

Author

Dr. Pierre Lamon
BlueBotics S.A.
PSE-C CH-1015 Lausanne
Switzerland
Email: pierre.lamon@bluebotics.com

ISBN 978-3-540-78286-5 ISBN 978-3-540-78287-2 (eBook)

DOI 10.1007/978-3-540-78287-2

Springer Tracts in Advanced Robotics ISSN 1610-7438

Library of Congress Control Number: 2008921245

Typesetting by the authors and Scientific Publishing Services Pvt. Ltd.

5 4 3 2 1 0

springer.com

STAR (Springer Tracts in Advanced Robotics) has been promoted under the auspices of EURON (European Robotics Research Network)

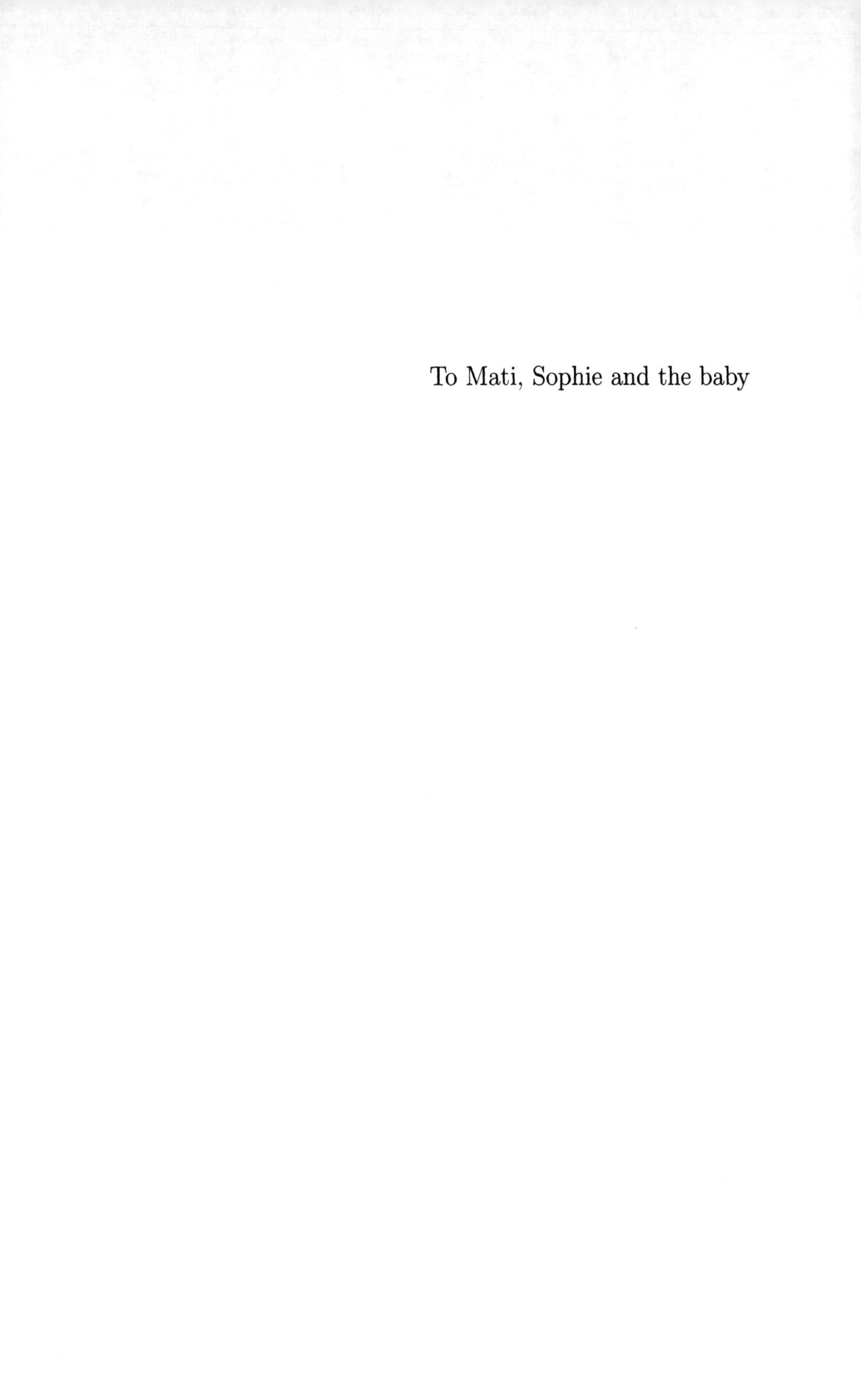

To Mati, Sophie and the baby

Foreword

At the dawn of the new millennium, robotics is undergoing a major transformation in scope and dimension. From a largely dominant industrial focus, robotics is rapidly expanding into the challenges of unstructured environments. Interacting with, assisting, serving, and exploring with humans, the emerging robots will increasingly touch people and their lives.

The goal of the new series of Springer Tracts in Advanced Robotics (STAR) is to bring, in a timely fashion, the latest advances and developments in robotics on the basis of their significance and quality. It is our hope that the wider dissemination of research developments will stimulate more exchanges and collaborations among the research community and contribute to further advancement of this rapidly growing field.

The monograph written by Pierre Lamon is the second the series devoted to tracking and control of robots in rough terrain. Research in this area has been mainly focused on 2D localization, on the assumption of flat surfaces. Whenever the rover has to climb over obstacles in cluttered environments, accurate 3D position tracking is crucial for both autonomous navigation and obstacle negotiation, as discussed in the introductory chapter. The trade-off between scientific and technical solutions makes this volume unique in the wide field of autonomous mobile robotics. The finalization of the approach on a real rover experimental platform allows a clear understanding of the influence of mechanical design, locomotion control and sensing on the pose tracking problem.

Remarkably, the monograph is based on the author's doctoral thesis, which received the prize of the Sixth Edition of the EURON Georges Giralt PhD Award devoted to the best PhD thesis in Robotics in Europe. A very fine addition to the Series!

Naples, Italy
December 2007

Bruno Siciliano
STAR Editor

Preface

During the first year of my doctoral thesis I went to Carnegie Mellon University for an internship. The goal was to implement the Mars Autonomy software[1] on Shrimp, a six wheeled rover with extended climbing capabilities. At that time, Shrimp had limited sensing capabilities and was only able to travel autonomously for about ten meters in a flat and simple environment. The robot couldn't reach farther goals because it rapidly lost track of its position. Thus, I decided to focus my research on position tracking in order to extend the range of autonomous navigation.

At that time, most of the research works in all-terrain rover navigation assumed flat environments and addressed 2D localization. Even though 2D localization is sufficient for many applications, the extension to 3D is necessary in cluttered environment, when the rover has to climb over the encountered obstacles. In such conditions, accurate 3D position tracking is crucial for both autonomous navigation and obstacle negotiation. The subject of my thesis and, *a fortiori* this book, was clear: 3D-position tracking and control for all-terrain robots.

Methodology

Autonomous mobile robotics is a fascinating field of research that involves technical and scientific domains. Thus, the research community working in this field is composed of people with very different backgrounds. One finds for example mathematicians, physicians, computer scientists, engineers and biologists. It is interesting to note that each researcher has his or her own definition of an autonomous mobile robot and has an individual way to address a given problem. However, two main categories of approaches can be distinguished i.e., the top-down and the bottom-up. One or the other is favored depending on the researcher's scientific and technical background. The top-down approach consists in developing a theoretical formulation of the problem and proposing a solution based on mathematical models. Although such a solution can be analyzed with respect to e.g. optimality and mathematical complexity, it does not necessarily

[1] The Mars Autonomy project: http://www.frc.ri.cmu.edu/projects/mars

work for the real application. Indeed, it often occurs that the models do not fully capture the reality or that they cannot be applied because they make use of unknown parameters. A failure to apply a model happens when the abstraction level is too high and when the technical constraints are not fully considered during the development phase. On the other hand, the bottom-up approach starts from a real application and proposes pragmatic solutions to given problems. The risk with such an approach is to tailor solutions in an incremental way, that is, to patch the system as problems arise. Such a reactive development favors the use of heuristics that may limit the system's performance and reliability.

The methodology used throughout this book is an attempt to reconcile the top-down and the bottom-up approaches and to avoid their respective traps. Even though the development was driven by the application, the bottom-up approach was not particularly favored. Thus, simple but valid heuristics were proposed only when modeling was not applicable. On the other hand, the technical constraints were considered during the modeling phases to avoid the generation of inapplicable models. In other words, the methodology used in this book tries to make the best tradeoff between scientific and technical solutions.

Acknowledgment

This work is an adventure during which I met many different people ready to help and to act as a source of inspiration. First of all, I am grateful to my advisor, Roland Siegwart, for convincing me to do a doctoral thesis at the Autonomous Systems Laboratory. All this has been possible thanks to his positive attitude, mentorship and support.

During the thesis, I had the chance to spend several months in other labs. Each time, the experience was very positive and stimulating. The first exchange was at CMU, were I discovered the world of linux and autonomy applied to rough terrain rovers. I would like to thank Reid Simmons for agreeing to supervise my work, Sanjiv Singh and Dennis Strelow for their help related to visual motion estimation, and Bart Nabbe and Jianbo Shi for their good advice. The next two exchanges took place at the Laboratoire d'Analyse et d'Architecture des Systèmes (LAAS-CNRS). In particular, I would like to thank Simon Lacroix, Anthony Mallet and Raja Chatila for their help and for having hosted me in excellent conditions.

Most of the student projects related to this research provided very good results, which helped to validate the theory through experiments with SOLERO. I would like to thank Ambroise Krebs for his excellent masters thesis, which enabled the development of a new approach to slip minimization in rough terrain. Also, I am grateful to Stéphane Michaud for the development of the mechanical structure of SOLERO, Martin Nyffenegger for the nice remote control interface, Benoît Dagon for the mechanical design of the panoramic vision system and Gabriel Paciotti for the stereovision rig.

The help of my colleagues was invaluable, and enabled me to develop the various systems discussed in this book. In particular, I would like to thank Grégoire Terrien and Michel Lauria for their expert advice related to the mechanical aspects, Agostino Martinelli for the mathematics, Daniel Burnier, Ralph Piguet and Gilles Caprari for the electronics and finally Rolf Jordi and Frédéric Pont for the questions related to informatics. The positive atmosphere in the lab provided favorable conditions for efficient and constructive work. Special thanks to Marie-José Pellaud, Nicola Tomatis, Daniel Burnier and my office-mate Gilles Caprari for their psychological support. I'm grateful to everybody in the lab for the great time I've spent during these four years.

Thanks also to the members of my thesis committee, Simon Lacroix, Paolo Fiorini and Bertrand Merminod, for their careful reading of the thesis and for their constructive feedback.

Lausanne, Switzerland
December 2007

Pierre Lamon

Contents

1 Introduction

1.1 Autonomy in Rough Terrain

Researchers have worked for years toward making mobile robots move by themselves and take their own decisions. Thanks to the efforts of a large research community and the evolution of technology, fully autonomous robots are today ready for applications in structured environments. However, their level of autonomy is still limited and the environments in which they are deployed are often engineered in order to guarantee reliability. Most successful applications are limited to indoor, office and industrial environments.

Outdoors, the operation of mobile robots is more complex, and a lot of effort has been deployed recently toward enabling a greater level of autonomy for outdoor vehicles. Such robots find their application in scientific exploration of hostile environments like deserts, volcanoes, the Antarctic, or other planets. There is also a high level of interest in such robots for search and rescue operations after natural or industrial disasters. Two examples of such applications are:

- The NASA project "Life in Atacama" aims to search autonomously for life in the Atacama desert in Chile. The first results are very promising but the following extract illustrates the difficulties encountered: "The farthest Zoë ran autonomously was 3.3 kilometers but on average a traverse would terminate after just over 200 meters" [2].
- The recent NASA "Mars Exploration Rover" mission (MER) aimed to understand how past water activity on Mars has influenced the red planet's environment over time. Because of the narrow bandwidth of communication and time delay, the tele-operation of the rovers was a slow process. For safety reasons, autonomous navigation was enabled only on relatively easy terrain and for short traverses.

These examples show that human supervision is still required for operating rovers in rough terrain and that further effort is required in order to enable fully autonomous operation.

P. Lamon: 3D-Position Track. & Cntrl. for All-Terrain Robots, STAR 43, pp. 1–5, 2008.
springerlink.com

1.2 The Open Challenges of Rough Terrain Navigation

The aim of this section is to describe the challenges when navigating in unknown and rough terrains. The main difficulties to cope with are: the lack of prior information, perception and locomotion.

1.2.1 Lack of Prior Information

There is a long-standing paradox between exploration and localization. In the past, navigators had to explore and map unknown regions while keeping track of their own position, which is difficult without a consistent map. For a mobile robot, the problem is the same when exploring a new area. There is no way to guarantee an optimal path between two points without any prior information about the encountered terrain topology and soil types. In order to reach a distant goal, the robot has to gain knowledge progressively about the explored environment, and store it in such a way that it can be used later for planning a path to the final destination. The hazards have to be detected and avoided as the robot explores new areas. This forces the robot to frequently replan its path, and for that, keep track of its own position.

1.2.2 Perception

Perception is a difficult task in mobile robotics because sensors are error-prone and their measurements are uncertain. In comparison with indoor environments, perception in natural scenes is even more complex and the acquired measurements are more difficult to analyze and interpret. For example, changing lighting conditions strongly affect the quality of acquired images and the vibrations due to uneven soils yield noisy signals.

In order to illustrate the problems involved in interpreting data, let us consider scans acquired by a 2D laser range finder with a 360° field-of-view. In an office environment, two scans taken from the same spot but with different heading contain the same information. A simple scan-matching algorithm confirms that both scans have been taken from the same position. On the other hand, a small heading change can lead to a large change in the rover's attitude in rough terrain. Even if the scans have been taken from the same position they can be completely different. This example demonstrates the substantial increase of perception complexity in 3D and shows the importance of choosing appropriate sensor configurations for outdoor environments.

Another problem of sensing in cluttered terrains lies in the limited field-of-view. The occlusions caused by the obstacles and terrain slope changes force the robot to adopt a complex motion strategy: the maneuvers shall maximize the information gain while limiting the risk of being trapped.

1.2.3 Locomotion

Indoors, the environment can be modeled as obstacle and obstacle-free areas and it is assumed that the robot can move freely within the obstacle-free regions.

In rough terrain, this simple representation is not possible because it is not straightforward to determine if a specific area is traversable. Many parameters such as the type of soil, the climbing performance of the rover and the terrain topology have to be considered. For example, a single rock of the size of the wheel's diameter can be overcome on a flat terrain, but can cause the rover to tilt over when encountered on a steep slope. Similarly, a flat and rocky patch could be easily crossed, whereas a sandy slope can represent an obstacle.

These examples illustrate the complexity of locomotion in rough terrains. Thus, both planning a safe path and controlling the rover actuators to execute the planned trajectory are difficult tasks. An optimal all-terrain locomotion concept together with a good wheel controller to optimize traction enables reaching more challenging areas and increases the performance of the system.

1.3 Research Context and Scope

A large part of the literature concerning autonomous navigation in rough terrain focuses on high-level functionalities such as environment modeling, perception and path planning. In [60], traversability maps are used instead of passable/impassable maps to plan a path through an unknown scene. The D* algorithm is used to dynamically replan a path as the robot acquires more information about the environment. The authors of [26] propose a unified process considering both perception planning and path planning so that the most relevant perception can be performed regarding the current goal of the robot. This provides a way to both optimally explore an environment while planning a path to the goal. Reference [40] presents the state-of-the-art on autonomous navigation in unknown terrains and proposes solutions to integrate the required functionalities in a consistent way. This work also insists on the importance of localization for autonomous navigation and the necessity to use a set of concurrent and complementary algorithms to produce robust position estimates. Reference [18] focuses on the selection of feasible trajectories based on the kinematic constraints of the rover and a digital-elevation map. Ensuring the proper execution of the generated trajectory is still an open challenge.

Considering the state-of-the-art and the open challenges of autonomous navigation in rough terrain, this book aims to contribute toward gaining a better understanding of the problems involved. In particular, it proposes concrete solutions for improving both locomotion and position tracking, which are the most limiting factors for autonomous navigation in rough terrain.

In this work, we are interested in applications, where no absolute sensing mechanism is available for long period of time and where the type of soil is unknown *a priori*. A good example of such application is the MER mission. During this mission, on-board rover's position tracking was primarily performed by the IMU, wheel-odometry, and sun-finding techniques. In cases where the rover experienced slippage caused by, e.g., traversing on loose soils or steep slopes, the onboard visual odometry technique was applied [22]. At the end of the Martian solar day (sol), telemetry data was used together with imagery in a bundle-adjustment algorithm to correct the positions of the rovers. The rovers were typically commanded only once per sol using a pre-scheduled sequence of specified commands (e.g., turn on the spot 30°, drive forward for 1 meter, etc.). So, having an accurate position estimate during the operation of the rovers was of critical importance [22]. The maximum range the rovers could reach within a sol was limited by the positioning error accumulated by dead reckoning. Because the position error grows as a function of the traveled distance, any method for reducing this error accumulation will considerably increase the traveled distance per sol.

Thus, the intent of this research work is to combine different approaches to reduce as much as possible the errors of dead reckoning. In particular, we show how 3D position tracking can be improved by considering rover locomotion in challenging environments as a holistic problem. Most of the research work focuses on a particular aspect of position tracking and/or is limited to relatively flat and smooth terrains. This work proposes to extend position tracking to environments comprising sharp edges such as steps and to consider not only sensing but also mechanical aspects (chassis configuration) and motion control (slip minimization).

In rough terrain, it is crucial to carefully select appropriate sensors and use their *complementarity* to produce reliable position estimates. Each sensor has its own advantages and drawbacks depending on the situation and thus relying on only one sensor can lead to catastrophic results. Although the choice of sensors is important, it is not the only aspect to consider. Indeed, the specific locomotion characteristics of the rover and the way it is driven also have a strong impact on pose estimation. The sensor signals might not be usable if an unadapted chassis and controller are used in rough terrain. For example, the signal-to-noise ratio is poor for an inertial measurement unit mounted on a four-wheel drive rover with stiff suspensions. Likewise, odometry provides bad estimates if the wheel controller does not include slip minimization, or if the kinematics of the rover is not taken into account in the algorithms. The aim of this research work is to reduce the position error growth by considering all these aspects together.

1.4 Structure of the Book

This book is structured as follows.

Chapter 2 presents the design of the rough terrain rover SOLERO capable of passively handling obstacles with sizes ranging up to two times the wheel diameter. This design has shown great potential in exploring hazardous environments. In the framework of this research, specific software tools and hardware have been developed for the prototype, making the system fully operational to run experiments in real conditions.

In Chapter 3, a new approach to compute 3D-motion increments based on wheel encoders and chassis state sensors is presented. It is called 3D-Odometry. Because the method accounts for the kinematics of the rover, it provides better results than the standard approach. When combined with an efficient mechanical structure such as SOLERO's, it allows producing accurate three dimensional motion estimates in rough terrain. This contributes toward more accurate pose tracking.

A physics-based controller minimizing wheel slip is proposed in Chapter 4. Minimizing wheel slip not only minimizes odometric errors (better position tracking), but also enhances the climbing performance of the rover (better locomotion). This controller is generalized and can be applied to any kind of passive mechanical structure with wheels.

Finally, Chapter 5 proposes an approach for combining proprioceptive and exteroceptive sensors to track the rover's position robustly in three dimensions. The results of sensor fusion integrating 3D-Odometry, inertial sensors and Visual Motion Estimation based on stereovision (VME) are presented. The experiments demonstrate how each sensor contributes to increase the accuracy and robustness of the 3D-motion estimation. The approach can accommodate any number of sensors, of any kind.

2 The SOLERO Rover

SOLERO (Solar-powered Exploration Rover) is the name of a rover built in a study carried out jointly by the Swiss Federal Institute of Technology of Lausanne (EPFL), Switzerland, and von Hoerner & Sulger GmbH (vH&S), Germany, under a contract from the European Space Agency (ESA). The objective of this activity was to develop a system design for a regional exploration rover, including breadboarding for the demonstration of locomotion capabilities, payload accommodation, power provision and control.

The most prominent feature of the SOLERO rover is its ability to operate relying on just-generated solar power, i.e., with minimal use of batteries (used only for contingency or peak power demands). A number of solar-powered vehicles operate in this mode on Earth. However, as Mars is farther from the sun than Earth, the power available from sunlight is significantly reduced and thus designing a working and reliable system is much more difficult. We have achieved power consumption optimization by combining smart power management and an efficient locomotion system. The calculations for power demand have been based on results obtained from ground testing of the breadboard rover and on accurate models of the Martian environment (i.e., reduced gravity, reduced intensity of sunlight, dust contamination on the solar arrays, temperatures, terrain surface, etc.). It was found that the rover should incorporate energy storage for peak demand and contingency situations only, allowing the solar array to be sized for smaller loads and resulting in a lighter, smaller and more reliable rover system. More information about the project can be found in [50, 16].

At the end of the project, one of the breadboards was significantly modified to serve as a research platform for autonomous navigation and enhanced control. The purpose of this chapter is to describe the mechanical design, the sensors and the control architecture of SOLERO.

2.1 Mechanical Design of SOLERO

Most locomotion concepts for all-terrain rovers are based on wheels, caterpillars or legs. *Walking machines* (e.g. Dante [11]) are well-adapted to unstructured environments as they can ensure stability in a wide range of situations. However,

P. Lamon: 3D-Position Track. & Cntrl. for All-Terrain Robots, STAR 43, pp. 7–19, 2008.
springerlink.com

they are mechanically complex and require a lot of control resources. On a planar surface, such locomotion concepts are slow and have high-power consumption in comparison to other concepts. *Caterpillar* robots demonstrate good off-road abilities thanks to their high stability and good traction capability (large footprint), but suffer high-friction losses between surface and caterpillars while turning. Treads offer good mechanical robustness and a large footprint, allowing the robot to move on various types of terrain. Examples of such robots are the Nanokhod [70] or TeleMAX [5]. *Wheeled rovers* are optimal for well-structured environments like roads or buildings. Off-road, their performance depends on the typical size of obstacles encountered. This holds for Sojourner [61], Micro 5 [39] and Rocky 7 [72], which can typically cross obstacles of their wheel size, if the friction coefficient is high enough. Extending the climbing abilities of a wheeled rover requires the use of a special strategy and implies dedicated actuators and a complex control procedure. Examples of active wheeled rovers are Octopus [43], Marsokhod [38], Hybtor [46], SpaceCat [42] and Nanorover [68]. A complete study of locomotion concepts based on wheels can be found in [44, 41].

The classification we generally use to study locomotion concepts differentiates between active and passive locomotion. Passive locomotion uses passive suspensions, having no sensors or additional actuators to guarantee stable movement, whereas an active robot employs a closed control loop to maintain stability during motion. Under this definition, Sojourner, Rocky 7 and Micro 5 are passive robots; Nanorover, SpaceCat, Octopus and walking machines are active robots; Marsokhod, Nanokhod and Hybtor are either active or passive, depending on their locomotion mode. Active locomotion extends the mobility of a robot but increases system complexity and needs extended control resources. Increasing the number of active degrees of freedom has a negative impact on the following characteristics:

- Sensing: more sensing capability is required for controlling the robot (e.g., encoders, force and torque sensors, etc.). This necessitates more cabling and increases the complexity of the overall architecture.
- Control: the control gets more complex as the number of degrees of freedom increases.
- Weight: the actuated degrees of freedom require motors, which add weight to the structure.
- Energy: the actuators and additional sensors for the joints increase power consumption.
- Reliability: the more complex, the lower the reliability.

Given these constraints, passive designs are generally preferred, especially for space applications, where power, weight and reliability are critical. New passive structures based on parallel kinematics have been proposed recently, i.e., Shrimp [25] and Crab [64]. Such concepts offer excellent climbing performance without suffering from the drawbacks of active designs.

The mechanical structure of the SOLERO rover is similar to that of Shrimp [58, 25, 59], an all-terrain rover developed at EPFL. SOLERO has one wheel mounted on a fork in the front, one wheel attached to the main body at the rear and two bogies on each side (Fig. 2.1).

Fig. 2.1. (**a**) SOLERO mechanical structure without sensors; (**b**) prototype equipped with a solar panel, power management board and scientific payload

Table 2.1. SOLERO main characteristics

Overall dimensions	$0.88 \times 0.6 \times 0.45$ m
Maximum speed	0.5 m/s^2
Rover's main body mass (inc. batteries, laptop etc.)	7.4 kg
Wheel mass (inc. motor, gears)	0.7 kg
Steering mechanism mass	0.6 kg
Spring constant	357 N/m
Wheel diameter	0.15 m
Ground clearance	0.208 m

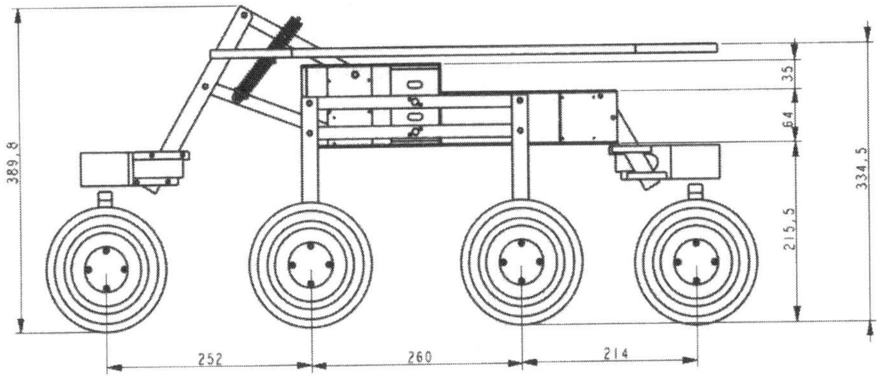

Fig. 2.2. Mechanical dimensions of SOLERO, left view (in *mm*)

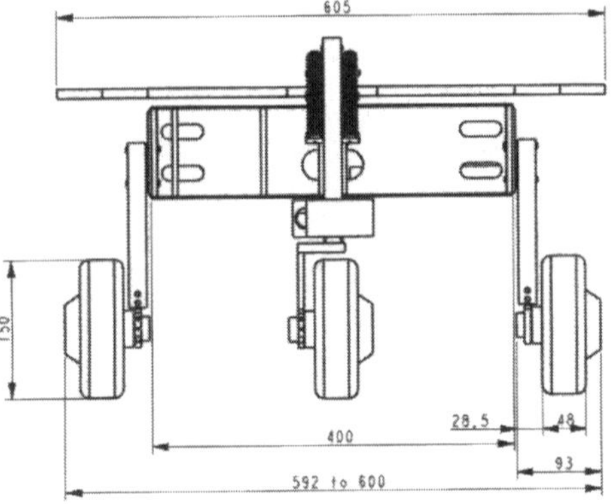

Fig. 2.3. Mechanical dimensions of SOLERO, front view (in mm)

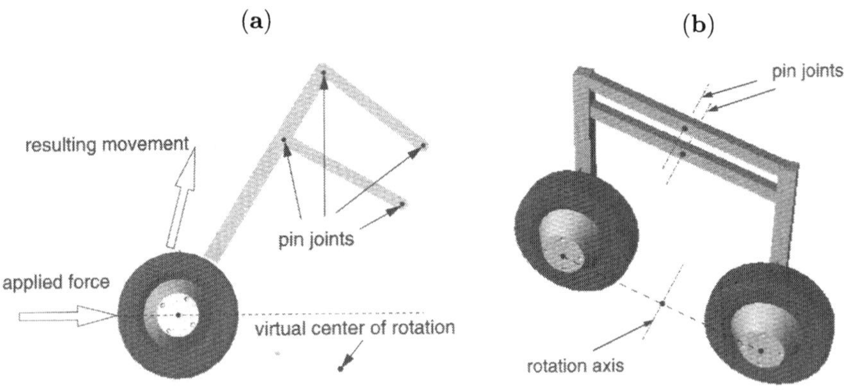

Fig. 2.4. Parallel mechanisms of SOLERO: (**a**) front fork kinematics. Because the instantaneous rotation center is placed below the wheel axis, the fork passively folds for climbing an obstacle; (**b**) parallel bogie with its virtual rotation axis placed between the wheels

The main characteristics of SOLERO are summarized in Table 2.1. Figures 2.2 and 2.3 depict the rover's overall mechanical dimensions. A more detailed model with all parameters is available in Appendix A.

The parallel architecture of the bogies and the spring-suspended fork provide a high ground clearance while keeping all six motorized wheels in ground contact at any time. This ensures excellent climbing capabilities over obstacles up to two times the wheel diameter, as well as excellent adaptation to all kinds of terrains.

The front fork has two functions: its spring suspension guarantees ground contact of all wheels, and its particular parallel mechanism produces a passive elevation of the front wheel if an obstacle is encountered. As depicted in Fig. 2.4a, the instantaneous rotation center is placed below the wheel axis. The result is that a horizontal force applied toward the wheel center translates into an elevation motion of the fork. The equations of the wheel center trajectory together with the fork parameters are presented in Appendix A.1.3

The bogies provide lateral stability. To ensure similarly good ground clearance and climbing capabilities, their virtual center of rotation is set to the height of the wheel axis using the parallel configuration shown on Fig. 2.4b.

Finally, the steering of the rover is realized by synchronizing the rotation of the front and rear wheels and the speed difference of the bogie wheels (skid-steering).

2.2 Control Architecture

The overall control architecture of SOLERO is presented in Fig. 2.5. The rectangular boxes represent computers and the rounded rectangles represent the sensors and actuators. SOLERO is equipped with two computers communicating through a crossover ethernet cable. The computer, called *solerovaio*, is a laptop in charge of image processing. It acquires images from a stereovision rig and an omnicam through a firewire (IEEE1394) bus, and transmits preprocessed data to the second computer, called *soleropc104*. This second computer has access to all the other sensors and actuators of the robot. It reads data from an inertial measurement unit through a RS232 port and interfaces an I²C bus using the parallel port. The devices attached to the I²C bus are: six wheel controllers, three servo-controllers, one angular sensor module (reading the three suspension angles) and an electronic board for the energy management of the rover. *soleropc104* acts also as a gateway for the rover's subnet. A host computer (*soleroap*) can connect to the subnet through a wireless ethernet interface. This allows, for example, the downloading of images, controlling the rover remotely through a graphical user interface and reading the rover state.

Fig. 2.6 presents the different sensors that have been mounted on SOLERO and the control system, including the actuators, the computational units and the electronic devices.

2.2.1 Sensors and Actuators

This section presents each sensor and actuator in more detail.

The I²C Modules

The use of an I²C bus allows for the easy integration of various types of sensors and actuators (up to 127 modules can be connected). Because each module contains its own processing unit, the computational load on the master processor is substantiality reduced; the main processor can read pre-processed data directly and send commands to the modules.

EPFL developed various I²C modules implementing interfaces for different kinds of sensors and actuators, including infrared and ultrasound distance sensors, linear camera, accelerometers, inclinometer, GPS, servo controller and DC motor controller. For SOLERO, two new types of I²C modules have been developed: an absolute angular sensor and an energy management module.

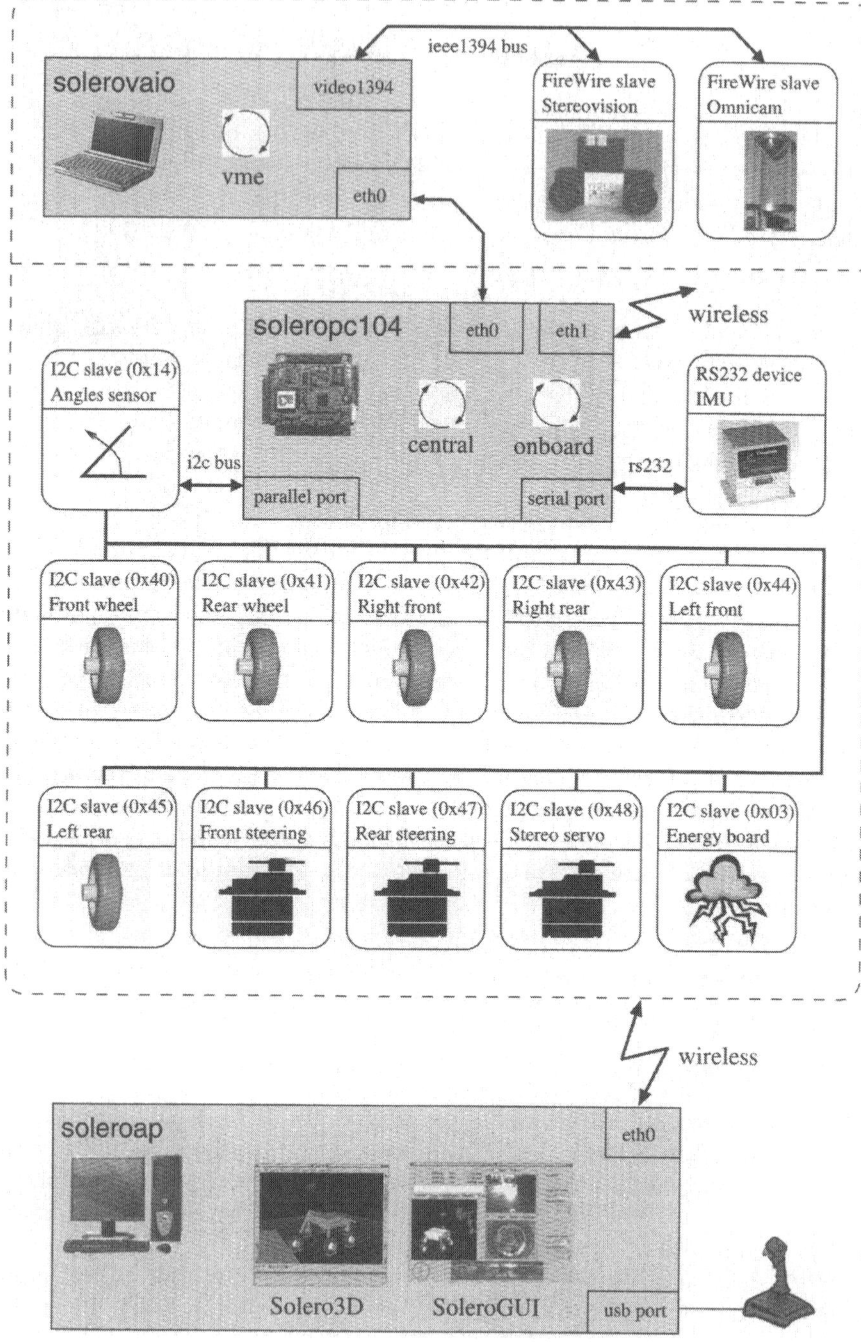

Fig. 2.5. Block diagram of the control architecture

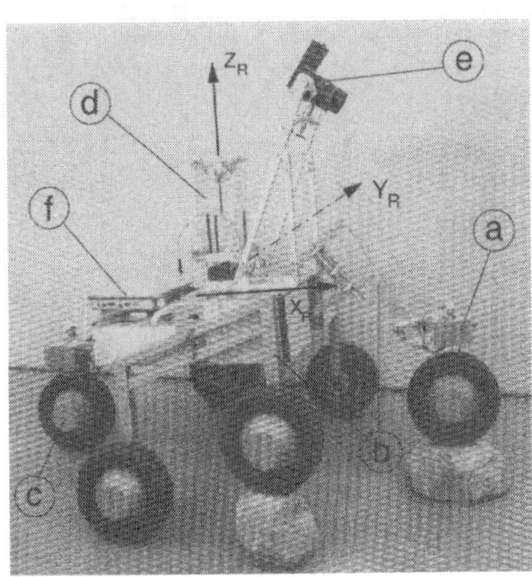

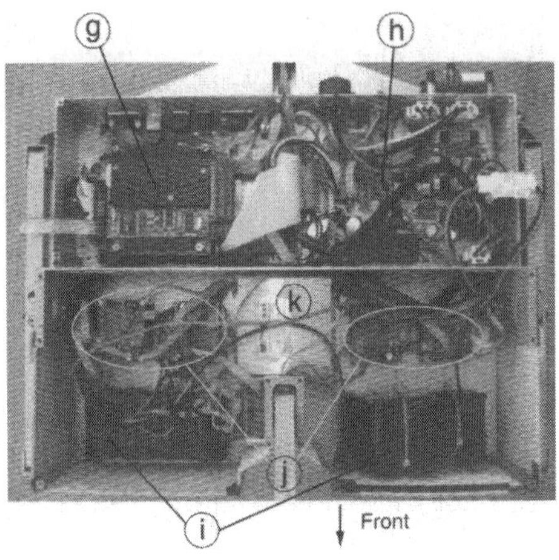

Fig. 2.6. Sensors, actuators and electronics of SOLERO: (**a**) steering servo mechanism; the same design is used for the rear wheel; (**b**) passively articulated bogie and spring-suspended front fork (equipped with absolute angular sensors); (**c**) 6 motorized wheels (DC motors); (**d**) omnidirectional vision system; (**e**) stereovision module, which can be oriented around the tilt axis; (**f**) laptop (solerovaio); (**g**) micro-computer (soleropc104); (**h**) energy management board; (**i**) batteries (NiMh 7000 mA h); (**j**) I²C modules (motor controllers, angular sensor module, servo controllers, etc.); (**k**) inertial measurement unit

Angular Sensor

In order to know the state of the chassis, the angular position of the fork and the bogies have to be measured. The angle of a pin joint is obtained by measuring the direction of the field of a magnet fixed to the rotating axle. The magnetic field direction is measured by means of a magneto-resistive bridge that is fixed to the main body. This contactless sensing mechanism is robust against drift, compact, and provides angle measurements with a precision of less than 0.2°. This accuracy is much higher than that obtained using a potentiometer-based approach. Furthermore, our solution provides absolute angles and does not require initialization each time the system is started (initialization would be required if a relative encoder were to be used). The components used to measure the front fork angle are depicted in Fig. 2.7.

Energy management board

The energy management board of SOLERO together with its physical interfaces are depicted in Fig. 2.8. Its main characteristics are summarized here:

- Two power supplies can be connected: a battery and an external power supply (DC voltage or solar panel in the range 11–15 V). An external switch or the I²C interface is used to switch between the two sources.
- Delivers regulated 5 V (10 A) and 12 V (1.2 A) for the system.
- Digital switch for the motors and system voltage.

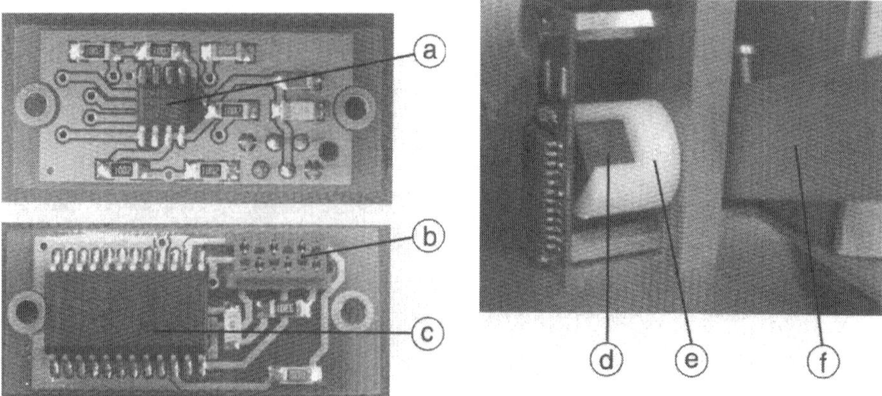

Fig. 2.7. Angular sensor for the front fork: (**a**) magneto-resistive bridge; (**b**) communication and power bus; (**c**) linearization chip; (**d**) magnet; (**e**) magnet holder (fixed to the axle); (**f**) lower arm of the front fork

- Currents and voltages of all the sources and consumers are measured and monitored.
- The battery voltage is monitored and the board warns the user by means of a blinking led and an acoustic signal. Below a certain voltage, the system is turned off automatically in order to protect the battery and the system.
- The battery status can be read from a seven-segment display.
- All the functions of the board can be accessed through I²C commands, i.e., switch on/off, battery status, currents and voltages, etc.

Stereovision

The stereovision rig is a MegaD module from VidereDesign, which can acquire grayscale images up to 1280×960 pixels at 15 Hz. The combination of 2/3" CMOS chips and lenses with 4.8 mm focal length results in a field-of-view of $85° \times 69°$. The rig is mounted on top of a mast and can be oriented around the tilt axis (Fig. 2.9). This allows the rig to keep ground features in the field-of-view of the cameras even if the rover is tilted upward or downward. In order to keep the center of gravity as low as possible, the motor is mounted close to the rover's main body. The rotational motion is transmitted to the stereovision module by means of a traction pole.

A lot of effort has been deployed to design a system with high stiffness and low mechanical play between the parts. This is important because the relative position between the camera coordinate system and the rover coordinate system has to be known precisely. Inaccuracies in the estimation of this transformation can lead to significant localization errors.

Omnicam

The omnicam depicted in Fig. 2.10 has been especially designed for SOLERO. Its main components are a camera and a panoramic mirror. A transparent cylinder

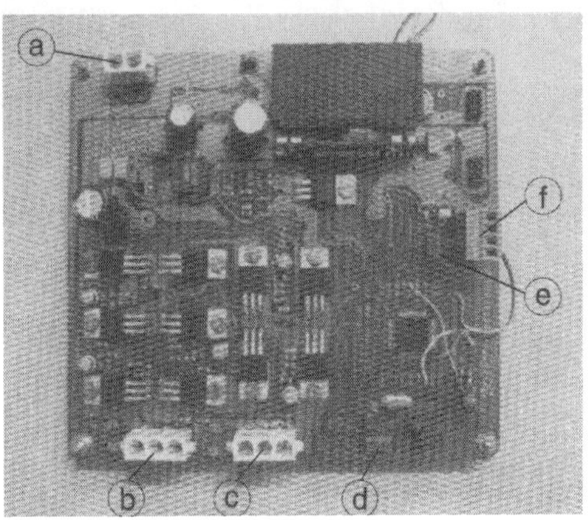

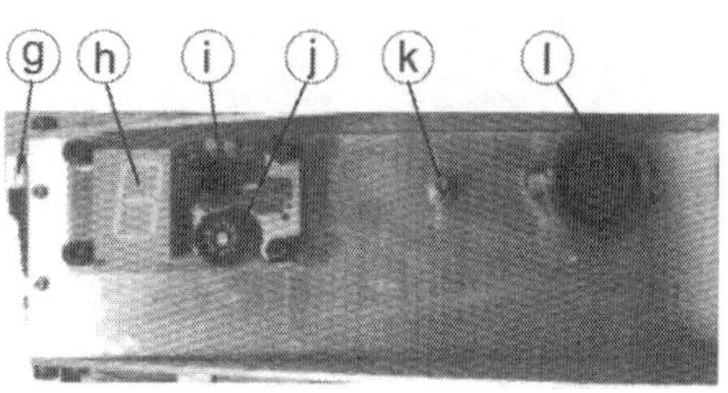

Fig. 2.8. The energy management board and the rear panel: (**a**) battery connector; (**b**) external charger connector; (**c**) external power supply connector; (**d**) I^2C bus connector; (**e**) external display connector; (**f**) regulated 5–12 V supplies; (**g**) source switch; (**h**) battery status display; (**i**) system state display; (**j**) low battery warning buzzer; (**k**) main system switch; (**l**) rear panel connector (used for both external power supply and charger)

(a) 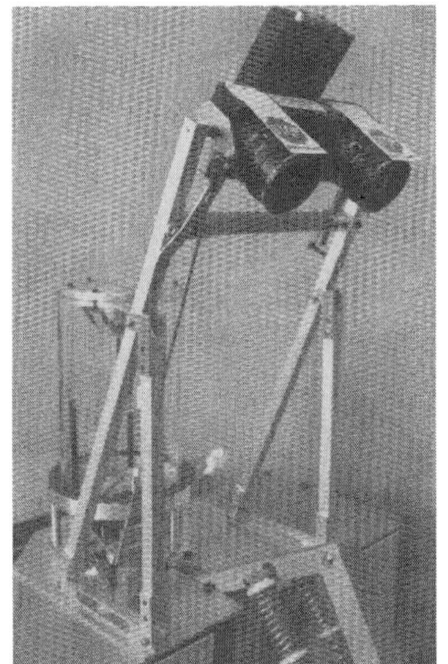

(b) 

Fig. 2.9. Stereovision support with tilt mechanism: (**a**) global view; (**b**) enlarged view of the motor transmission. The motor is equipped with an optical encoder, used for measuring the tilt angle.

Fig. 2.10. Omnicam hardware with a transparent cylinder for dust protection

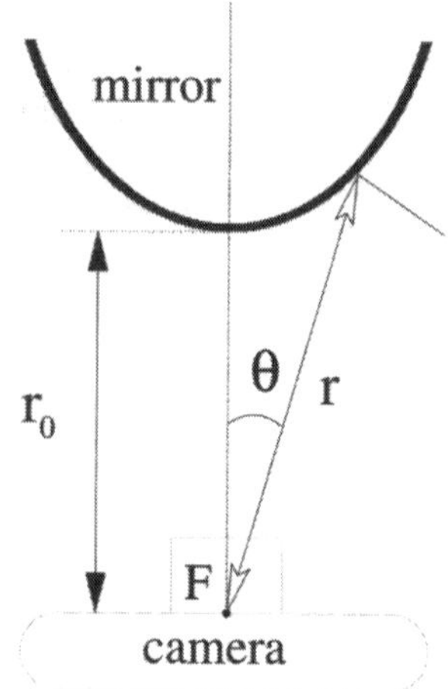

Fig. 2.11. Equiangular mirror and vision system parameters

is used to hold the mirror and to protect the vision system against dust. The camera has been chosen for its compactness and low-power consumption. These properties make it suitable for mobile applications. Grayscale and color images up to 640 × 480 pixels can be acquired through firewire at a rate of 20 Hz. The characteristic equation of the mirror is given by

$$\cos\left(\theta\frac{1+\alpha}{2}\right) = \left(\frac{r}{r_0}\right)^{-\frac{1+\alpha}{2}} \tag{2.1}$$

with r_0 the distance from the focal point F to the mirror, α the intrinsic parameter of the mirror, θ the angle of the ray, and r the distance from F to the mirror along the ray (see Fig. 2.11). Such a mirror belongs to the family of equiangular mirrors: each image pixel covers the same solid radial angle. Thus, when moving radially in the image, the shape of an image feature is less distorted than it would be by using other mirror shapes (such as spherical or parabolical). This facilitates feature tracking and data association between two consecutive images. More information about this kind of mirror can be found in [21, 52].

Inertial Measurement Unit

The Inertial Measurement Unit (IMU) is the VG400CC-200 device from Crossbow. It is a solid-state inertial measurement system that utilizes MEMS micromachined sensing technology. It is composed of a triad of accelerometers (velocity rate sensors) and gyroscopes (angular rate sensors), which are combined internally to provide roll and pitch angles in static and dynamic conditions (using a Kalman filter). Furthermore, the calibrated angular rates and accelerations are available and come together with a timestamp. This timing information allows us to perform accurate integration of angular and velocity rates over time.

2.2.2 Software Architecture

The system is divided into five functional modules running as separate processes, i.e., *vme*, *central*, *onboard*, *Solero3D* and *SoleroGUI* (see Fig. 2.5). The modules run on different computers and communicate using the Inter-Process Communication messaging system [1]. This IPC library, developed at Carnegie Mellon University, can send and receive complex data structures transparently, including lists and variable length arrays, using both anonymous "publish/subscribe" and "client/server" message-passing paradigms. In the current configuration, *vme* runs on *solerovaio*, *onboard* and *central* on *soleropc104*, and *Solero3D* and *SoleroGUI* on the remote computer *soleroap*. However, the architecture can be easily modified to accommodate another hardware configuration. For example, *vme*, *central* and *onboard* could all run on *solerovaio*.

Because *vme* and *onboard* exchange time critical data, the internal clocks of *solerovaio* and *soleropc104* are synchronized. This synchronization is guaranteed by Network Time Protocol [3] deamons running on both computers.

central acts as a server for the IPC network. It is responsible for routing the messages and holds the systemwide information (such as defined message prototypes). *onboard*, the main program of our software architecture, has access to the low-level sensors and actuators, i.e., the IMU and the I²C modules. Its main tasks are to perform sensor fusion and execute the motion commands sent by the remote control interface *SoleroGUI*. On *solerovaio*, the *vme* module has access to the stereovision and the omnicam through the firewire bus. After image

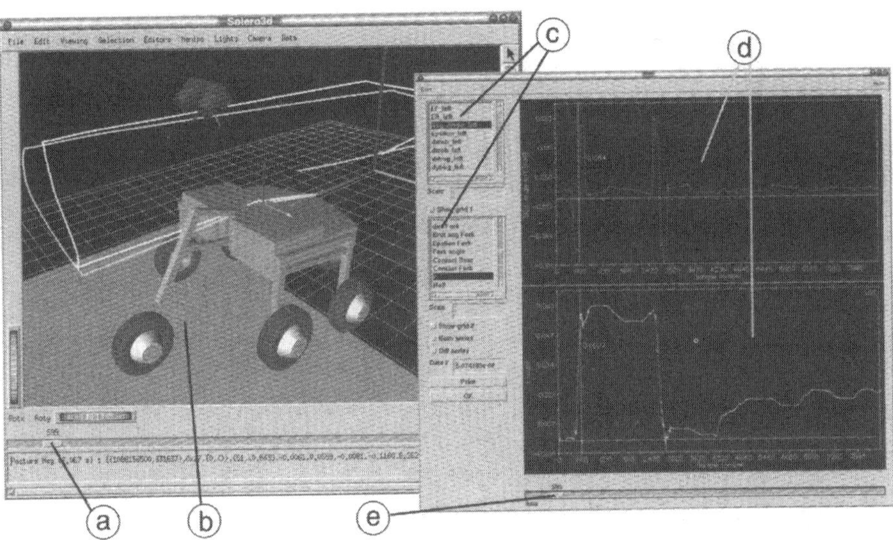

Fig. 2.12. The main window (*left*) and the data browser (*right*): (**a**) data replay slider. By manipulating this slider the scene is updated with the corresponding robot state; (**b**) 3D rendering area; (**c**) variable selection lists one and two; (**d**) plot areas one and two; (**e**) data browser slider

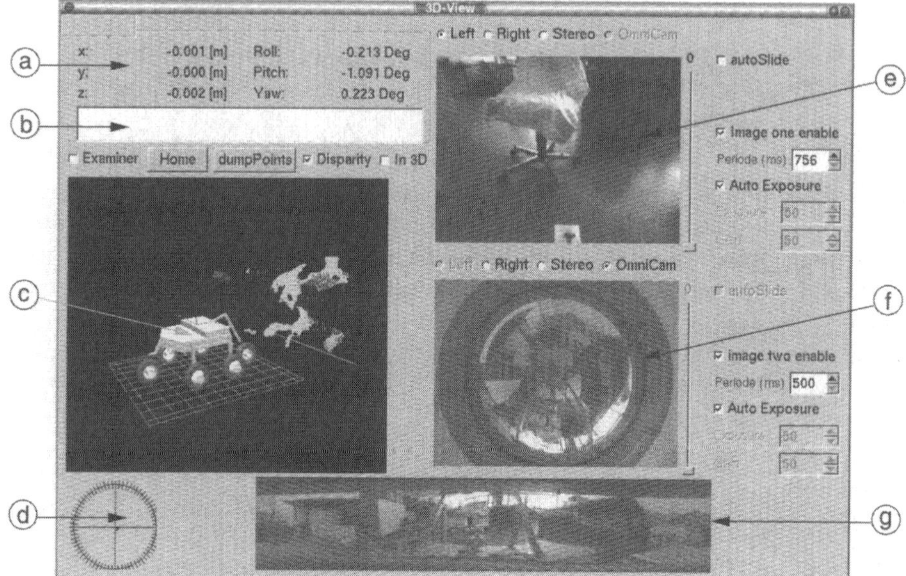

Fig. 2.13. The remote control interface for SOLERO: (**a**) robot's position and attitude; (**b**) text box for warning messages, e.g., the pitch angle exceeds a predefined value; (**c**) 3D representation of the robot state and 3D cloud of stereo points. The operator can interactively change the perspective; (**d**) rover control area. The motion commands are given by clicking and moving the mouse cursor. The robot can be also driven with a joystick; (**e**) first image area. The user can select the left, right, stereo or omnicam image. All the imagers' settings can be modified; (**f**) second image area; (**g**) panoramic view area.

acquisition, it performs image processing and sends the results to *onboard* for sensor fusion. The two remaining modules *Solero3D* and *SoleroGUI* are described in more detail in the following sections.

Solero3D

This program has been developed for visualizing and logging data acquired by the robot during an experiment. Its main use is for testing and debugging algorithms offline. Figure 2.12 shows the main window (*left*) and the data browser (*right*).

The main window integrates a full 3D rendering area, allowing the user to change views. By manipulating a slider, the set of data stored during an experiment can be replayed step by step. All the variables such as the robot position, the internal links' angles and the attitude angles are plotted in the data browser. This module is an important tool to test the system, debug and compare algorithm performance.

Remote Control Interface

A dedicated module called *SoleroGUI* has been developed for the tele-operation of SOLERO (Fig. 2.13). In order to ease remote control, the graphical user interface displays the images taken onboard the rover together with a 3D view of the current rover state. Furthermore, stereoscopic information is displayed in the 3D scene, allowing the operator to have a better understanding of the environment in front of the rover and thus properly avoid hazardous areas. A panoramic image is displayed at the bottom of the main window. It provides a wide field-of-view and helps the operator to navigate through the scene. Finally, all parameters of the imagers are accessible through dialog boxes.

Other features of the GUI (Graphical User Interface) are listed here:

- Warning messages such as "low battery" and "dangerous rover posture" are printed on the screen in order to avoid critical situations, which could damage the rover.
- The operator can control the rover with a game pad or the mouse. Smooth trajectories are generated using nonlinear optimization accounting for both maximal wheel acceleration and speed.
- Two operation modes are available. The first one is called "coordinated and allows the robot to drive on any circular arc. The second mode is called "non-coordinated" and only straight line and point-turning motion are allowed.
- A watchdog timer is implemented to detect communication problems. The GUI sends a signal every second and the rover stops in case the signal is not received.
- The GUI uses cross-platform libraries and can be compiled and run on different OS.

2.3 Summary

Because no generic hardware setup exists for such rovers, a lot of effort had to be deployed to make SOLERO a well-performing platform for research. Furthermore, a set of powerful tools has been developed for speeding up the process of debugging the algorithms and analyzing the data stored during the experiments. The modularity and portability of the system allows easy adaptation of new actuators and sensors. For example, a GPS can be easily added to the I^2C bus and more cameras attached to the firewire or USB buses. The possibility to access all the schematics and firmware of the I^2C sensors allows to have a low-level control of data transmission timings and thus improve the reactivity of the system. Off the shelf components are rarely well documented, especially concerning time-stamping of data, which is of high importance in robotics. The energetic autonomy of the system running on batteries depends on the intensity and duration of the driving phases. In average, the autonomy is around three hours, allowing to run long sessions of experiments. This autonomy can be doubled by replacing NiMh batteries by LiPo, while keeping the same weight.

3 3D-Odometry

Until recently, autonomous mobile robots were mostly designed to run in indoor environments that are partly structured and flat. Many problems arise in rough terrain and position tracking is more difficult. Although odometry is widely used indoors (2D), its application is limited in natural environments (3D). The wheels are more likely to slip because of the rough structure of the soil, and the error in the position estimation can grow quickly. For these reasons, odometry is generally avoided in challenging terrains. However, we can look at the problem differently and ask: "How could we limit wheel slip in rough terrain ?"

Two different aspects can be improved to minimize wheel slip. The first one is to develop a controller that guarantees good balance of wheel torques and speeds to optimize the robot's motion. A torque controller minimizing slip and maximizing traction is presented in Chapter 4. The second aspect is to improve the mechanical structure of the robot. Indeed, a well-designed mechanical structure yields a smooth trajectory and thus enables limited wheel slip. As described in Chapter 2, SOLERO can adapt passively to a large range of obstacles and allows limited wheel slip in comparison with rigid structures such as four-wheel drive rovers. Thus, the odometric information is usable even in rough terrain. A new technique called 3D-Odometry, which provides 3D motion estimates of SOLERO, is presented in this chapter.

3.1 3D-Odometry

Odometry is widely used to track the position (x, y) and the heading ψ of a robot in a plane π [19]. The vector $[x\ y\ \psi]_\pi^T$ is updated by integrating small motion increments sensed by the wheel encoders between two time-steps. When combined with the robot's attitude angles ϕ (roll) and θ (pitch), this technique allows the estimation of the six degrees of freedom in a global coordinate system, i.e., $[x\ y\ z\ \phi\ \theta\ \psi]_W^T$. The orientation of the plane π on which the robot moves is determined using an inclinometer and the motion increment is obtained by projecting the robot displacements measured in the plane π into the global coordinate system. This method, used in [40], is referred later as the *standard*

P. Lamon: 3D-Position Track. & Cntrl. for All-Terrain Robots, STAR 43, pp. 21–32, 2008.
springerlink.com

method. The approach works well only if the environment does not comprise too many slope discontinuities. Indeed, the pose estimation error grows quickly during transitions because it is assumed that the rover is locally moving in a plane. In rough terrain, this assumption is not appropriate and the transitions problem must be addressed properly. Odometry based on vehicle kinematics and wheel encoders are addressed in [34, 9]. However, the presented results focus mainly on straight trajectories and smooth terrains. In this work, we present an approach that deals with sharp obstacles and full 3D trajectories.

The 3D-Odometry computation is divided into two steps: a) the computation of the left and right sides of the robot (Sect. 3.1.1), and b) the computation of the resulting 3D displacement of the robot's center (Sect. 3.1.2). The main reference frames and variables used for 3D-Odometry are defined in Fig. 3.1.

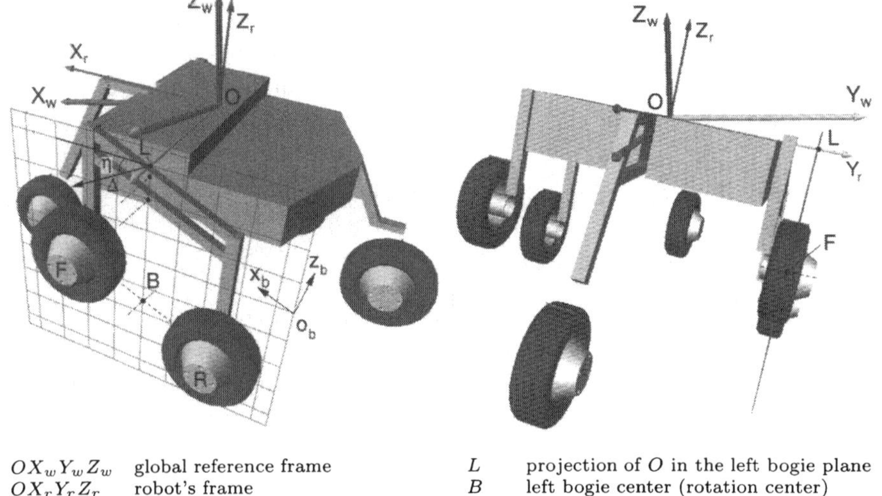

$OX_wY_wZ_w$	global reference frame	L	projection of O in the left bogie plane
$OX_rY_rZ_r$	robot's frame	B	left bogie center (rotation center)
O_bxz	bogie frame (in the bogie plane)	Δ, η	norm/direction of L's displacement

Fig. 3.1. Main variables and reference frames for 3D-Odometry

3.1.1 Bogie Displacement

For SOLERO, the motion increments of the left and right bogies are used to compute the 3D motion of the robot's center O. The aim of this section is to describe how to compute the displacement (Δ and η) of one bogie knowing the translations of the wheels (encoder data ER, EF) and the change of the bogie angle (ε) between state t and $t+1$ (see Figs. 3.2 and 3.3). In what follows, the equations are developed for the left bogie. However, the same method is applied for the right bogie and the corresponding equations are obtained using simple variables and parameter substitution.

For computing the displacement of L we proceed in two steps: a) The displacement of B is computed in the bogie's frame $O_b xz$ (Fig. 3.2) b) This motion increment is propagated through the bogie's mechanical structure to compute the effective displacement of L, expressed in the robot's frame OX_rZ_r (Fig. 3.3). The equations to calculate the displacement of B are similar to [31]. However, we extend the approach to calculate the displacement of a full parallel bogie (Δ and η). Because the distance between the wheels remains constant, one can write the following equations (ER, EF and ε being the parameters)

$$\begin{aligned} \overrightarrow{RF} &= \overrightarrow{RR'} + \overrightarrow{R'F'} + \overrightarrow{F'F} \\ \begin{pmatrix} h_b \\ 0 \end{pmatrix} &= \begin{pmatrix} ER\cos\rho_w \\ -ER\sin\rho_w \end{pmatrix} + \begin{pmatrix} h_b\cos\varepsilon \\ -h_b\sin\varepsilon \end{pmatrix} + \begin{pmatrix} -EF\cos\phi_w \\ EF\sin\phi_w \end{pmatrix} \end{aligned} \tag{3.1}$$

These equations can be solved for ϕ_w and ρ_w, which are the wheel-ground contact angles. However, this equation system can be inconsistent in some pathological cases. For example, if ε equals zero then ER must be equal to EF because the distance between the wheels remains constant. In practice, ER and EF can be slightly different because the wheels are subject to slip. When the equation

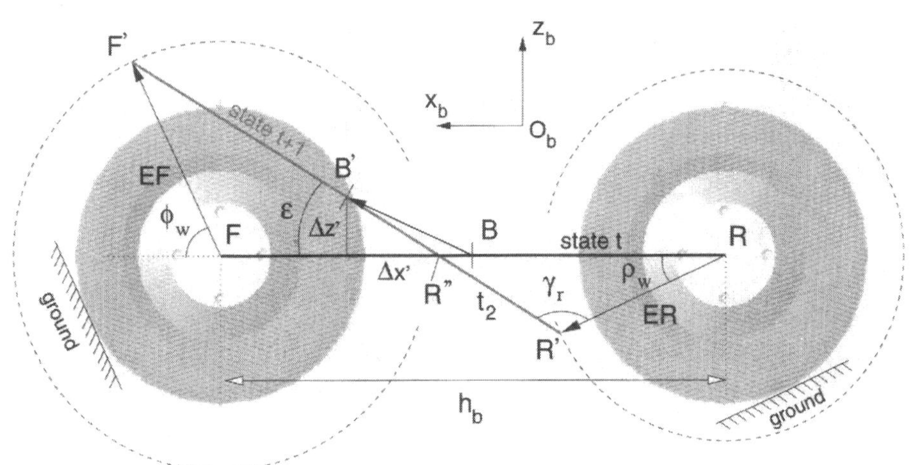

ER, EF	rear/front wheel displacement
ρ_w, ϕ_w	rear/front wheel's direction of motion
R, F	initial position of the rear/front wheel center
R', F'	final position of the rear/front wheel center
B, B'	initial/final position of the bogie center
ε	bogie's angular change
h_b	distance between the wheel centers
$\Delta x', \Delta z'$	x/z components of vector $\overrightarrow{BB'}$
$t1$	distance BR'' (not displayed on the figure)
$t2$	distance $R'R''$

Fig. 3.2. Displacement of B between state t and $t+1$. The final position of the rear/front wheel is on a circle of radius ER/EF centered at R/F, respectively.

system is inconsistent, we set the bogie displacement as being equal to the average of the displacements of the two wheels. In the general case, valid solutions are found and the sine theorem is applied in the $RR'R''$ triangle to compute $\Delta x'$ and $\Delta z'$, which are the coordinates of the displacement of B expressed in the bogie's coordinate system $O_b xz$

$$\Delta x' = t_1 + \left(\frac{h_b}{2} - t_2\right) \cos \varepsilon \qquad \qquad \Delta z' = -\left(\frac{h_b}{2} - t_2\right) \sin \varepsilon$$

with

$$t_1 = \frac{ER\,|\sin(\pi - \rho_w - \varepsilon)|}{|\sin \varepsilon|} - \frac{h_b}{2} \qquad \text{and} \qquad t_2 = \frac{ER\,|\sin \rho_w|}{|\sin \varepsilon|}$$

Fig. 3.3 defines the parameters for computing the displacement of L considering the displacement of B and the mechanical structure of the bogie. The effective bogie angle change between state t and $t+1$ is given by

$$\varepsilon = \theta_2 + \phi_2 - (\theta_1 + \phi_1) \tag{3.2}$$

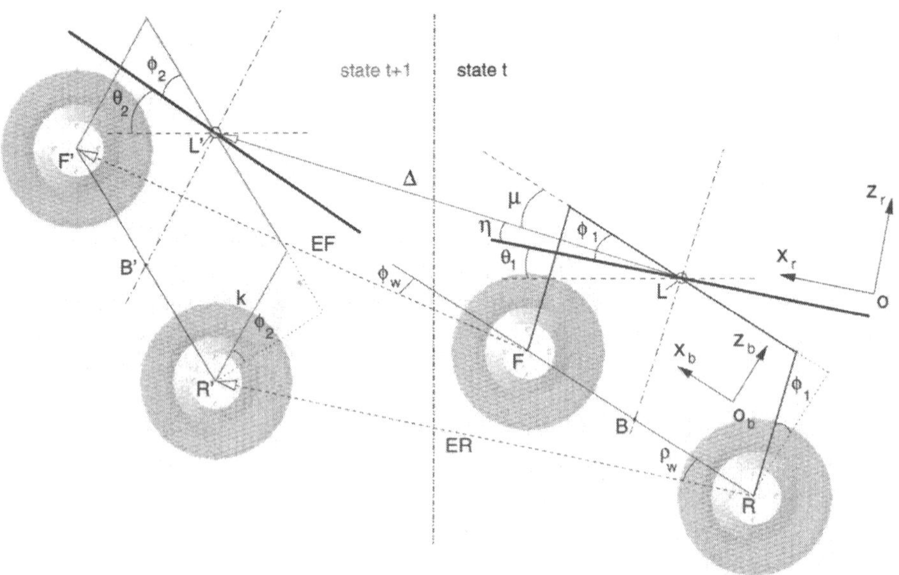

L	projection of O in the left bogie plane	μ	angle of LL' expressed in $O_b xz$
L'	final position of L (at time $t+1$)	η	angle of LL' expressed in $OX_r Z_r$
Δ	norm of LL'	θ_1, θ_2	initial/final pitch angle
$\Delta x, \Delta z$	coordinates of LL' expressed in $O_b xz$	ϕ_1, ϕ_2	initial/final bogie angle
k	bogie leg length		

Fig. 3.3. Bogie's configuration change between two time steps. The distance between the states has been artificially increased for the purpose of clarity.

Because the relative position of L with respect to B depends on the bogie configuration, the displacement of B and L are not the same. This effect is taken into account to compute the effective displacement of L. Considering that the angular changes and the motion increments between t and $t+1$ are small, the incremental corrections are given by

$$\begin{aligned} c_x &= -k\left(\sin\phi_2 - \sin\phi_1\right) \\ c_z &= k\left(\cos\phi_2 - \cos\phi_1\right) \end{aligned} \tag{3.3}$$

Then c_x and c_z are added to $\Delta x'$, $\Delta z'$ to compute the actual displacements of point L expressed in the bogie's coordinate system $O_b xz$

$$\Delta x = \Delta x' + c_x \tag{3.4}$$

$$\Delta z = \Delta z' + c_z \tag{3.5}$$

Finally, the norm of the displacement Δ and the motion angle η expressed in the robot's frame are given by

$$\Delta = \sqrt{\Delta x^2 + \Delta z^2} \qquad \eta = \phi_1 + \mu$$

where μ is defined as

$$\mu = -\arctan\left(\frac{\Delta z}{\Delta x}\right) \tag{3.6}$$

3.1.2 3D Displacement

The previous section described the computation of the norm Δ and the direction of motion η of one bogie. The aim of this section is to derive the equations for computing the 3D displacement of the robot's center O using the left and right bogie translations. The main schematic for the 3D-Odometry is depicted in Fig. 3.4. In the latter, the subscripts l and r are used to denote variables related to the left and right bogies respectively. The angles η_r and η_l define the tilt of the planes π_r and π_l containing C' and L'. C' and L' are situated on circles centered in C and L with radius Δ_r and Δ_l, respectively. These constraints lead to the following equations

$$\mathbf{n}_r \cdot \overrightarrow{CC'} = 0 \tag{3.7}$$

$$\mathbf{n}_l \cdot \overrightarrow{LL'} = 0 \tag{3.8}$$

$$\Delta_r^2 = x_r^2 + \left(y_r + \frac{W_b}{2}\right)^2 + z_r^2 \tag{3.9}$$

$$\Delta_l^2 = x_l^2 + \left(y_l - \frac{W_b}{2}\right)^2 + z_l^2 \tag{3.10}$$

The infinitesimal displacements of the left and right sides of the robot mainly occur in the bogie planes. When the robot is turning, however, the norm of the displacement of one side is larger than the other and that forces a fraction of the motion to occur out of the bogie planes (along Y_r). Because of the nonholonomic constraint, this displacement cannot be measured directly. Thus, we assume that the smallest displacement among Δ_r and Δ_l takes place in the corresponding bogie plane, giving to the other side an additional degree of freedom along Y_r. In the

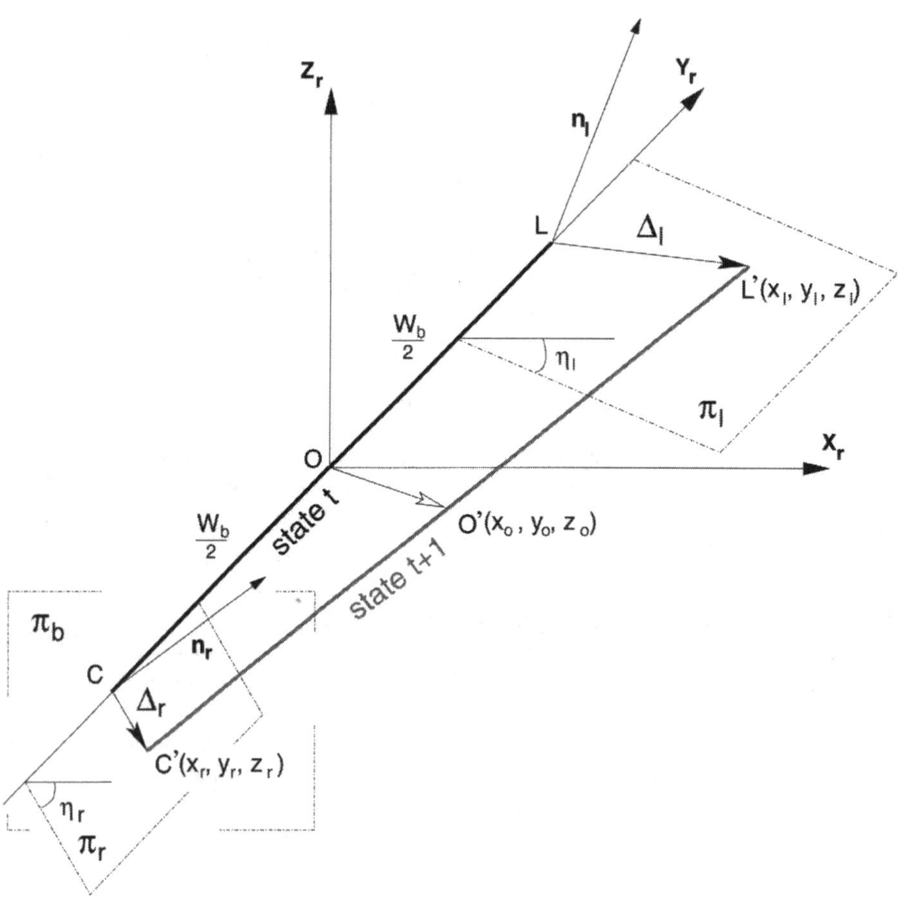

W_b	distance between the bogie planes	Δ_r, Δ_l	right/left absolute displacement
C, C'	initial/final position of the right bogie	π_r, π_l	right/left planes
L, L'	initial/final position of the left bogie	$\mathbf{n}_r$, $\mathbf{n}_l$	normal vectors of π_r π_l
O, O'	initial/final position of the robot center	π_b	plane // to OX_rZ_r, containing C
η_r, η_l	right/left displacement angles		

Fig. 3.4. 3D-Odometry, variables definition

example of Fig. 3.4, Δ_r is smaller than Δ_l. Therefore the right displacement vector is constrained to remain in the bogie plane π_b. This constraint is expressed by

$$\left(\frac{W_b}{2}\right)^2 + \Delta_r^2 = x_r^2 + y_r^2 + z_r^2 \tag{3.11}$$

In case Δ_r is bigger than Δ_l, one substitutes the subscript r with l in 3.11. Because the wheelbase W_b remains constant, one can write the additional constraint

$$W_b^2 = (x_l - x_r)^2 + (y_l - y_r)^2 + (z_l - z_r)^2 \tag{3.12}$$

Finally, the rover's displacement vector is obtained using

$$\overrightarrow{OO'} = \overrightarrow{OC'} + \frac{\overrightarrow{C'L'}}{2} \tag{3.13}$$

Solving the system of nine equations with nine unknowns formed by equation 3.7 to 3.13 leads to the solutions for $\overrightarrow{OC'}$, $\overrightarrow{OL'}$ and $\overrightarrow{OO'}$ (the nine unknowns). The yaw angle increment is computed using

$$\mathrm{d}\psi = \frac{x_r - x_l}{W_b} \tag{3.14}$$

The roll increment $\mathrm{d}\phi$ can be computed by substituting x_r and x_l by z_r and z_l in 3.14. However, we have chosen to rely on the value of the roll angle provided by the inclinometer because this is an absolute angle and therefore it is not subject to drift.

3.1.3 Contact Angles Estimation

The 3D-Odometry provides an estimation of the motion increment in three dimensions and the heading change of the rover. It is interesting to note that the contact angles of the bogie wheels are also computed by this method (see Fig. 3.2). The contact angles of the fork and the rear wheels can be computed using parameters substitution in 3.1. To estimate the rear-wheel contact angle, ε is replaced by the pitch change of the rover ($\mathrm{d}\theta$) and the norm of the robot's motion, computed by the 3D-Odometry, is used instead of EF. The same kind of parameter substitution is used to compute the contact angle of the front wheel. The wheel-ground contact angles are required for predictive wheel controllers for minimizing wheel slip. Such a controller is presented in Chapter 4.

3.2 Experimental Results

The Shrimp breadboard (see Fig. 3.5) has been used to compare the performance of 3D-Odometry and the standard method. In order to facilitate the estimation of the ground-truth, we used obstacles of known shape: this allows computation of the true trajectory of the rover using its kinematic model and the geometric dimensions of the obstacles. The rover was remotely controlled several times over the obstacles and the trajectories were computed online with both 3D-Odometry and the standard method. For each run, we measured the rover's final position and computed the absolute and relative errors. We tested the system for different obstacle configurations such as depicted in Figs. 3.6, 3.7 and 3.9. The obstacles are made of wood and the ground is a mixture of concrete and small stones.

Fig. 3.5. The Shrimp breadboard. This robot is a small scale version of SOLERO.

For all experiments, the position estimation error of the standard method grows quickly. This is due to the fact that the method does not account for the kinematic model of the robot and only considers the attitude of the main body. Whereas the consequences of this approximation are less relevant for smooth terrains (Fig. 3.6), they become disastrous while climbing sharp-shaped obstacles (Fig. 3.7).

On the other hand, the 3D-Odometry estimates the angle ε (3.2), which corresponds to the actual motion direction of the bogie center B. This angle, together with the wheel encoders' data allows computation of the wheel-ground contact angles and the norm and direction of motion of a bogie (point L or C). This direction corresponds neither to the pitch of the main body nor to the mechanical angle of the bogie: it is the actual direction of motion of the point L or C in the global coordinate system. Thus, the motion increment Δz (3.5) is computed and used to correct the trajectory along z_r during the slope transitions. The corrections corresponding to the first experiment are plotted in Fig. 3.8. They are positive when the first wheel starts climbing the obstacle, zero when there is no slope change and become negative when the bogie finally reaches the top of the obstacle. The slope transitions are well detected and this allows correction of

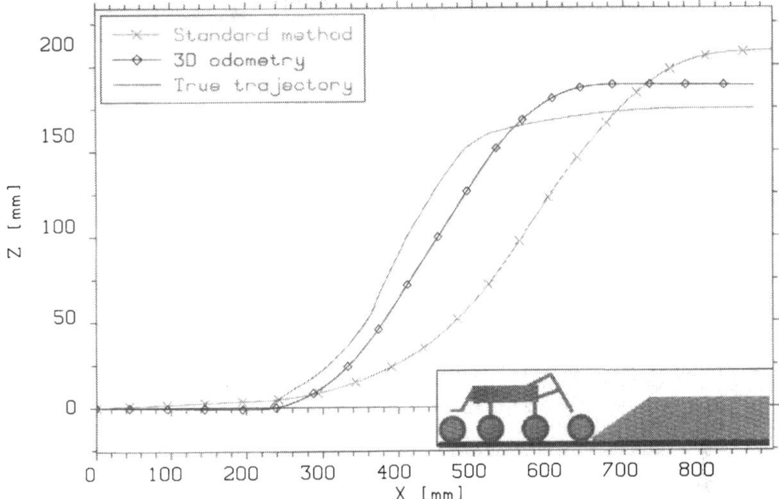

Fig. 3.6. Steep slope experimental setup. The robot starts in front of the obstacle, climbs a 35°slope and stops on top after traveling a distance of 870 mm along the x direction. The height of the obstacle is 175 mm. The slope changes are relatively smooth for this experiment.

the z-coordinate to thus better track the robot's position. The position estimations computed by 3D-Odometry are more accurate and the numerical figures presented in Tables 3.2 and 3.2 confirm these qualitative results.

In order to correctly analyze the error sources of the 3D-Odometry, the main mechanical parameters of the robot were calibrated before the experiments. That way, the remaining errors are only due to nonsystematic errors and errors introduced by our approach. The wheel diameters and the wheelbase of the robot have been calibrated using [19]. The pitch offset of the IMU has been estimated using the trajectory computed by the standard approach on flat ground. After the robot completes a full loop, the final height shall be equal to the initial height. The pitch offset of the IMU, O_{imu}, is obtained using

$$O_{imu} = \arcsin\left(\Delta h/s\right) \tag{3.15}$$

where Δh is the error accumulated along the z axis and s the total path length. After the calibration of the offsets, the remaining errors related to the 3D-Odometry shall be due to nonsystematic errors and approximation of the algorithm. The first source of error we might think about is wheel slip. In case of slip the calculated distance is bigger than the measured one and the results presented in Tables 3.2 and 3.2 can be interpreted that way. However, wheel slip is not the biggest source of error in these experiments. The errors are mainly along the z direction and they are due to nonlinearities, variation of the wheel

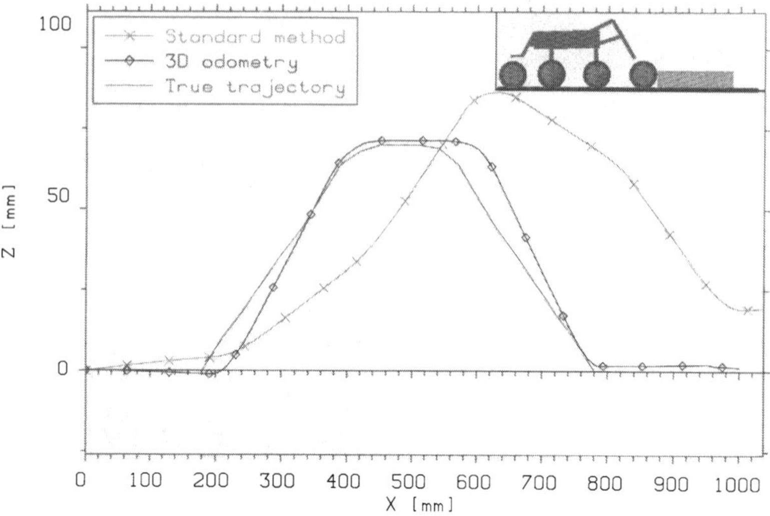

Fig. 3.7. Sharp edges experimental setup. The robot goes over a 300 by 70 mm obstacle and stops after 1 m travel. Even with sharp slope transitions, the 3D-Odometry trajectory accurately follows the obstacle shape.

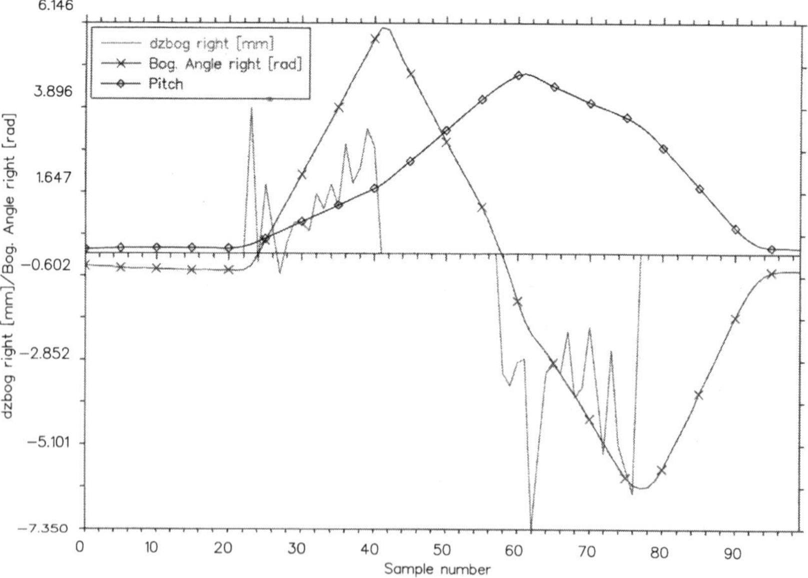

Fig. 3.8. z-coordinate correction for the first experiment. For clarity, both Pitch and Bogie Angle curves have been scaled by a factor of 12 and 8, respectively. The time-step between two samples is about 60 ms (computation time for the 3D-Odometry).

Table 3.1. Relative error for the steep slope experiment (870 mm)

Measured		3D-Odometry		Standard	
x	y	x	y	x	y
864	175	871	188	896	209
873	175	876	187	904	210
872	175	877	186	905	209
875	175	878	185	908	208
870	175	873	186	903	208
Mean error		0.5%	6.4%	3.7%	19.2%

Table 3.2. Relative error for the sharp edges experiment (1000 mm)

Measured		3D-Odometry		Standard	
x	y	x	y	x	y
993	0	1000	1	1038	19
1010	0	1012	5	1056	24
1015	0	1008	7	1062	25
1009	0	1008	6	1059	25
996	0	1002	4	1046	22
Mean error		0.2%	2.7%	5.5%	13.2%

diameters, and inaccuracy in the mechanical dimension. For the steep slope experiment, the final averaged error of 11 mm can be explained by a remaining angular offset. Indeed, an offset of 1°leads to an error of around 15 mm in the z direction for an 870-mm horizontal motion. For the sharp edges experiment, these errors canceled out because of the symmetry of the obstacle.

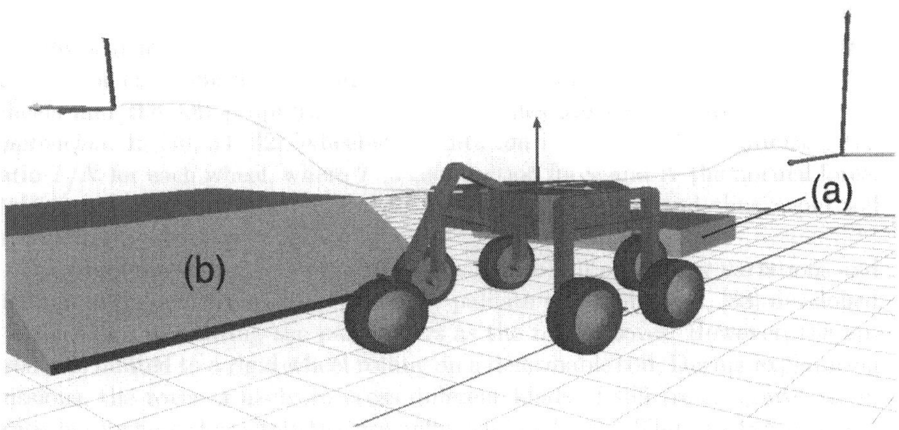

Fig. 3.9. Full 3D experiment performed with the Shrimp. Only the right bogie wheels climbed the obstacle (**a**). Then, the rover was driven over obstacle (**b**) (with an incident angle of approximatively 20°).

The experimental setup used to test the full 3D capability of the 3D-Odometry is depicted in Fig. 3.9. Like for the first two experiments, the real robot was used. This time, the rover followed the sequence of preprogrammed commands consisting of a) going straight for 1 m b) turning to the right 70° c) go straight for 1 m. Only the right bogie wheels climbed the first obstacle (**a**) whereas the other wheels kept ground contact. Then, the rover was driven over the second obstacle (**b**) with an incident angle of approximatively 20°. The value of such an experiment is that it forced the chassis to adopt asymmetric configurations and allowed us to test the full 3D-capability of the 3D-Odometry. The true final position and orientation of the rover was hand-measured and compared with the computed final position. The average error at the goal was only $\epsilon_x = 0.02\,m, \epsilon_y = 0.02\,m, \epsilon_z = 0.005\,m, \epsilon_\psi = 3°$ for a total path length of around $2\,m$. This corresponds to a relative error of 1.4, 2, 2.8 and 4%, respectively.

3.3 Summary

This chapter described a new approach called 3D-Odometry that estimates the 3D displacement of a rover in rough terrain. The sensory inputs are the wheel encoders of the four bogie wheels, the bogie angular sensors and an inclinometer. It has been shown that this approach performs better than the standard method used traditionally. The position estimation is significantly improved when the rover overcomes sharp-shaped obstacles because the method accounts for the kinematic model of the rover.

Thanks to its nonhyperstatic mechanical structure, SOLERO overcomes obstacles in a smooth manner, with limited wheel slip. When applying 3D-Odometry, such a design enables the use of odometry as a means to estimate the rover motion in rough terrain. Moreover, the accuracy of odometry can be further improved using a controller minimizing wheel slip. The description of such a controller is presented in the next chapter.

4 Control in Rough-Terrain

For wheeled rovers, motion optimization is generally related to minimizing wheel slip. Minimizing slip not only limits odometric error but also increases the robot's climbing performance and traction. Different approaches to slip minimization in rough terrain can be found in the literature.

The controller developed in [69] derives from the Anti-lock Breaking System (ABS) and uses the information of wheel slip to correct individual wheel speed. Reference [13] proposes a velocity synchronization algorithm, which minimizes the effect of the wheels "fighting" against each other. The first step of the approach consists in detecting which of the wheels are deviating significantly from the nominal velocity profile. Then a voting scheme is used to compute the required velocity set point change for each individual wheel. Because such methods adapt the wheel speeds when slip has already occurred, they are referred to as *reactive approaches*.

A controller providing better performance might be developed by considering the physical model of the rover and wheel-soil interaction models. Thus, the traction of each wheel is optimized considering the load distribution on the wheels and the soil properties. Such approaches are referred to as *predictive approaches*. In [30, 34, 32], wheel-slip limitation is obtained by minimizing the ratio T/N for each wheel, where T is the traction force and N the normal force. Reference [74] proposes a method to minimize slip ratios and thus avoid soil failure due to excessive traction.

The predictive approaches are very sensitive to soil parameter variations and difficult to implement on real rovers. To palliate this limitation, [33] developed a method for estimating the parameters as the robot moves. However, the approach is limited to a rigid wheel rolling on a deformable soil. During exploration missions, the rover is likely to cross different kinds of soil (rock, gravel, sand, grass, etc.) whose characteristics are unknown in advance. Thus, to apply model-based approaches, both the wheel-soil interaction models and parameters shall be adapted depending on the encountered terrain type. This requires developing a terrain classifier able to detect all kinds of soil and a way of estimating the parameters for each class as the robot moves. In practice, it is tedious to

P. Lamon: 3D-Position Track. & Cntrl. for All-Terrain Robots, STAR 43, pp. 33–51, 2008.
springerlink.com

implement because it is impossible to cover the entire range of terrain types. In case *a priori* knowledge about the environment is available, a method such as presented in [20] might be implemented to classify and detect a limited set of terrain classes.

In this chapter, we propose a predictive controller that accounts for the load distribution on the wheels and that does not require *a priori* knowledge about wheel-soil interaction models and terrain classification. The proposed closed-loop approach estimates the rolling resistance torque online using the sensing capabilities of the rover instead of relying on possibly inaccurate or unknown models.

4.1 Quasi-static Model of a Wheeled Rover

The speed of an autonomous rover is limited in rough terrain because the navigation algorithms are computationally expensive (image processing, path planning, etc.) and the onboard processing power is limited. In this range of speeds, typically smaller than 20 cm/s, the dynamic forces can be neglected and a quasi-static model is appropriate. Such a model can be solved for contact forces and motor torques knowing the state of the robot and the wheel-ground contact angles. To develop such a model, a mobility analysis of the rover's mechanical structure is performed (Sect. 4.1.1). The analysis produces a consistent physical model with the appropriate degrees of freedom at each joint. Then the forces are introduced and the equilibrium equations are written for each part composing the rover's chassis (Sect. 4.1.2).

4.1.1 Mobility Analysis

The mobility of a rolling robot in straight motion should ideally be one, indicating that the robot can move in a constrained direction. Grubler's Mobility Equation in three dimensions [47], also known as the Kutz–Bach Criterion, is written as

$$MO = 6n - 5f_1 - 4f_2 - 3f_3 - 2f_4 - f_5 \tag{4.1}$$

where n is the number of mechanical parts of the system and f_j the number of joints of each type ($j = 1 \dots 5$). For example f_1 is the number of pin joints and f_3 the number of spherical joints. The mobility equation is a guideline for determining if a system is statically determinate. Many real systems contain redundancy in links and joints resulting in hyperstatism. A four-legged table, for example, is statically indeterminate if considered rigid. More sophisticated modeling methods are required to analyze the distribution of forces in a hyperstatic system. Another approach is to model selective joints with additional degrees of freedom. Intelligent selection of these joints can minimize the error associated with a quasi-static solution. While the modeled kinematic chain is a simplification, it can be good enough to support motor control.

Mobility analysis of SOLERO

In a first step, one can consider the wheel-ground contacts as spherical joints and all the pin joints in the mechanism as one degree of freedom (DOF) revolute. For the SOLERO, the calculation of the mobility using 4.1 is -20 rather than 1. The system is, therefore, significantly hyperstatic and requires a modified model for a possible quasi-static solution. Two significant modifications to joint degrees of freedom assist the model. The first one involves the representation of the wheel-ground joint mobility. For a standard wheel without slip, the joint that represents the wheel-ground contact can be modeled as a spherical joint allowing three degrees of freedom (rotations about the three axes). Motor torque on the wheels will directly affect the forces in that contact plane. Lateral forces are not influenced by the motor torque. Therefore, the system was modeled with the lateral forces being carried by the wheel fixed to the body (rear wheel) and the wheel on the front fork. The wheels on the bogies were modeled with no resistance in the lateral direction (4 degrees of freedom). We have chosen such a model because not enough information exists to assess how the lateral forces are distributed among all the wheels and because the error due to the simplification has almost no influence on the controller. The second modification acts on the representation of the redundant kinematic chains. It is possible to model selected joints on redundant kinematic chains with more degrees of freedom. This results in forces being transmitted through direct flow patterns. Because the model is being used to optimize motor torques, inaccuracies in the internal linkage forces have minimal effect. Figure 4.1 shows the resulting kinematic model of SOLERO.

The numbers at the link connections indicate the degrees of freedom of that joint. The resulting model is mechanically equivalent to the real structure. The

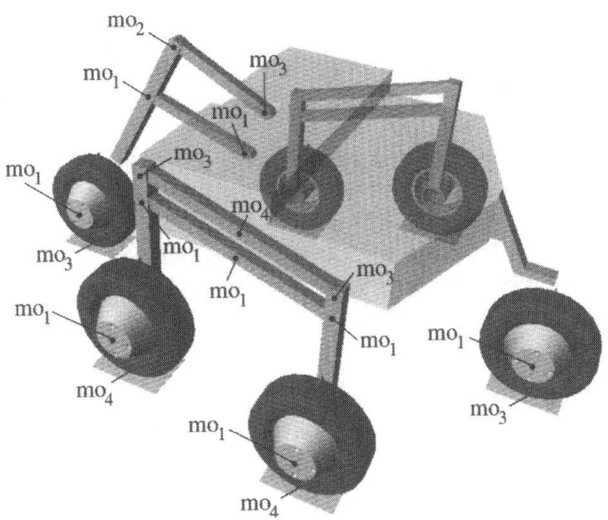

Fig. 4.1. Joints mobility for SOLERO

mobility of the bogies and the fork is one and the final mobility can be calculated using 4.1 to produce

$$MO = 6 \times 18 - 5 \times 14 - 4 \times 1 - 3 \times 7 - 2 \times 6 = 1 \tag{4.2}$$

4.1.2 A 3D Static Model

For a 3D static model, 6 equations (3 torques and 3 forces) are applied to each body, containing ground reaction forces, gravity forces (weight) and external forces. Dynamic forces are considered to be negligible because of the low speed. The mobility analysis is used to introduce the right number of forces/torques at each joint. In 3D, the number of generalized forces (n_g) to introduce follows the rule

$$n_g = 6 - mo \tag{4.3}$$

where mo is the mobility of the joint. For example, five generalized forces are introduced for a pin joint ($mo = 1$): only one rotation is free, while all the translations (3 forces) and the remaining rotations (2 torques) are blocked.

Model of SOLERO

SOLERO has 18 parts and is characterized by $6 \times 18 = 108$ independent equations describing the static equilibrium of each part and involving 14 external ground forces, 6 internal wheel torques and 93 internal forces and torques for a total of 113 unknowns. The weight of the fork and the bogies' link has been neglected, whereas the weight of the main body and the wheels is considered. Of course, it is possible to reduce this set of independent equations because we have no interest in implicitly calculating the internal forces of the system. The variables of interest are the three ground contact forces on the front and the rear wheels, the two ground contact forces on each wheel of the bogies and the six wheel torques. This makes 20 unknowns of interest and the system can be reduced to $20 - (113 - 108) = 15$ equations. This leads to the following matrix equation

$$M_{15\times 20}\,\mathbf{U}_{20\times 1} = \mathbf{R}_{15\times 1} \tag{4.4}$$

where M is the model matrix depending on the geometric parameters and the state of the robot, $\mathbf{U}$ a vector containing the unknowns and $\mathbf{R}$ a constant vector. The details of the model together with the mechanical parameters are described in Appendix A.

4.2 Torque Optimization

It is interesting to note that there are more unknowns than equations in 4.4. That means that the set of wheel-torques guaranteeing the static equilibrium is infinite. This becomes obvious when considering that one motorized wheel is enough to make the robot move. This characteristic can be used to control the traction of each wheel and to select, among all the possibilities, the set of torques minimizing slip. Other functions, such as energy, can be used instead. In this chapter, we focus on slip minimization and this section describes the concepts and the optimization algorithms.

4.2.1 Wheel Slip Model

The intent of this section is to formulate a holistic model of a robot to control the wheel motor torques in order to minimize wheel slip. Therefore, it is helpful to review the governing equations on wheel slip and explain which function should be minimized to reach this goal. The model presented in what follows assumes a rule of proportionality between traction and the normal force on the wheel: the more pressure on the wheel, the more traction it can carry before slipping. This proportionality rule is not perfectly verified in all circumstances. However, such a model is valid in most cases and is appropriate because we are not interested in exactly computing the forces at the interface but in minimizing wheel slip. Figure 4.2 shows the common forces acting on the wheel of a mobile robot.

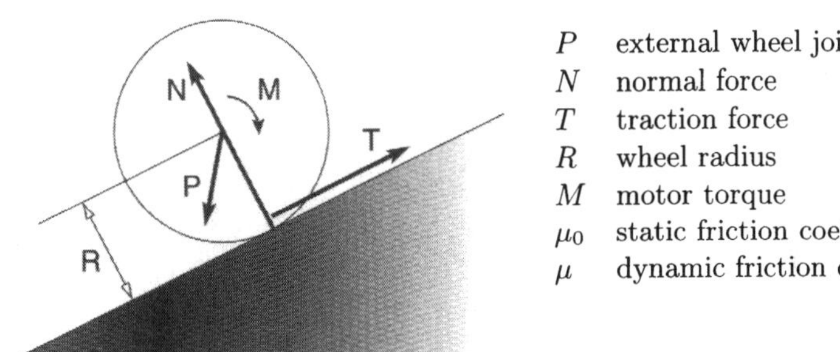

P	external wheel joint force
N	normal force
T	traction force
R	wheel radius
M	motor torque
μ_0	static friction coefficient
μ	dynamic friction coefficient

Fig. 4.2. Acting forces on a wheel and friction coefficients definition

The wheel is balanced if the friction force fulfils 4.5. This case represents static friction. If the static friction force cannot balance the system, the wheel slips and the friction force is set by 4.6.

$$F_{static} \leq \mu_0 N \qquad (4.5) \qquad\qquad F_{dynamic} = \mu N \qquad (4.6)$$

In order to avoid wheel slip, the friction force that depends directly on the motor torque M should satisfy

$$T = \frac{M}{R} \leq \mu_0 N \tag{4.7}$$

The above equations suggest that there are two ways to reduce wheel slip. First, assume that μ_0 is known and set

$$T \leq \mu_0 N \tag{4.8}$$

In fact, it is difficult to know μ_0 precisely because it depends on the kind of wheel-soil interaction. During exploration, the kind of soil interacting with the wheels is not known, which makes μ_0 impossible to predetermine. Another way to avoid wheel slip is to first assume that the wheel does not slip. It is then possible to calculate the forces T and N as a function of the torque and the result is optimized in order to minimize the ratio T/N. Accounting for this assumption

$$\frac{T}{N} = \frac{\mu_n N}{N} = \mu_n \tag{4.9}$$

μ_n is similar to a friction coefficient. In minimizing this ratio, then minimizing μ_n, we optimize our chances that this coefficient is smaller than the real friction coefficient μ_0. If this is the truth, no slip occurs. Therefore, it is possible to minimize the ratio T/N without knowing the real static friction coefficient. The second method is used here, because it is more robust.

4.2.2 Optimization Algorithm

The controllable inputs of the system are the six wheel torques. Since there are five more unknowns than equations, it is possible to write an equation expressing the torques as linearly dependent (a proof is presented in Appendix A.2.1). The 14 other equations define the external forces as a function of the torques. The model of SOLERO is indeterminate because there are fewer equations than variables and the set of solutions is five-dimensional (number of wheels - 1). The goal of the optimization is to minimize slip and this can be achieved by maximizing the traction forces, which is equivalent to minimizing the function

$$f = max(T_i/N_i) = max(\mu_i) \tag{4.10}$$

with $i = 1 \dots 6$. The minimum of f is denoted μ_m in what follows. μ_m can be seen as the minimal-needed friction coefficient that guarantees no slip. Because it is difficult to do reasoning in five dimensions, a simpler 2D robot referred to as ThreeWheels (see Fig. 4.3) is used to present our optimization algorithm. This process is then extended for the complete model of SOLERO.

The model of the ThreeWheels rover has nine unknowns, i.e., two forces and one torque on each wheel and seven equations: three global equations, one torque equation for each wheel and one torque equation for the fork. That leads to a two-dimensional solution space. Equation 4.11 expresses the torque of the first wheel as a function of m_2 and m_3 (m_1, m_2 and m_3 are linearly dependant) and 4.12 and 4.13 express the normal and tangential forces, respectively. $\alpha, \beta, \gamma, \epsilon$ and δ, are parameters that depend on the rover's state and geometry (see Appendix A.2).

$$m_1 = \epsilon_1 \, m_2 + \epsilon_2 \, m_3 + \delta \tag{4.11}$$

$$N_i = \alpha_{1i} \, m_2 + \beta_{1i} \, m_3 + \gamma_{1i} \tag{4.12}$$

$$T_i = \alpha_{2i} \, m_2 + \beta_{2i} \, m_3 + \gamma_{2i} \tag{4.13}$$

with $i = 1 \dots 3$

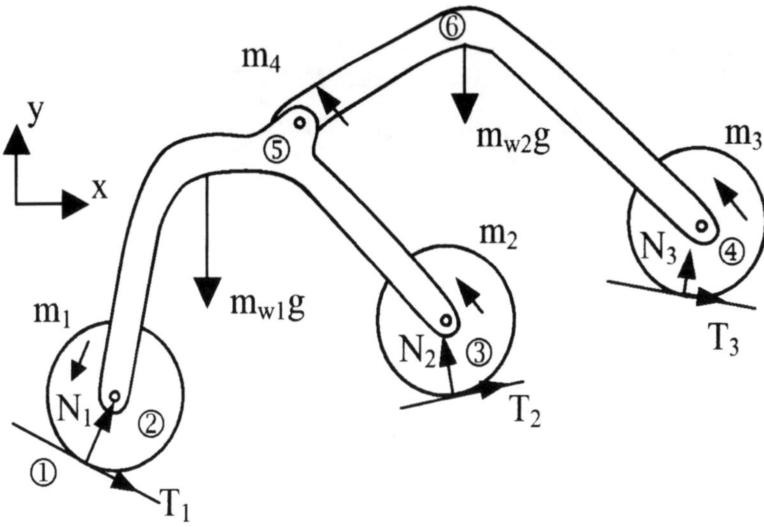

Fig. 4.3. The ThreeWheels 2D model. This rover belongs to the passively suspended robots family. m_4 is an uncontrolled torque generated by a torsion spring with known characteristics.

The optimal solution is found by minimizing the function f, which is plotted in Fig. 4.4

$$f(m_2, m_3) = \max\left(\frac{T_1}{N_1}, \frac{T_2}{N_2}, \frac{T_3}{N_3}\right) = \max(\mu_1, \mu_2, \mu_3) \tag{4.14}$$

The optimization problem is nonlinear because the functions μ_1, μ_2 and μ_3 are hyperbolic. Our optimization method, whose scheme is presented in Fig. 4.5, uses a combination of several modules. Each module is activated alternatively depending on the result of the previous module in the chain. The first step of the optimization process consists of initializing the algorithm with a set of torques that satisfy the model. A set of equal torques is chosen as the initial solution because it always verifies the model (see Appendix A.2.2 for the proof). Then, this solution is checked against the physical constraints:

a. positive normal force: the normal forces N_i must be greater than zero (the asymptotes of the hyperbolic functions define the sign inversion limit)
b. motor torque saturation: the wheel torques m_i must be smaller than the maximal motor torque $MaxTrq$.

If this solution is valid, it is taken as the initial solution for the fixed-point optimization (**A**). If it does not fulfill the constraints, a valid initial solution is computed using the simplex method (**B**). The optimal solution is then provided either by (**A**) or the gradient optimization (**C**). We have chosen this scheme because most of the states are handled by (**A**), which is computationally very light in comparison with a standard monolithic, constrained optimization algorithm. The different modules are now presented in more detail.

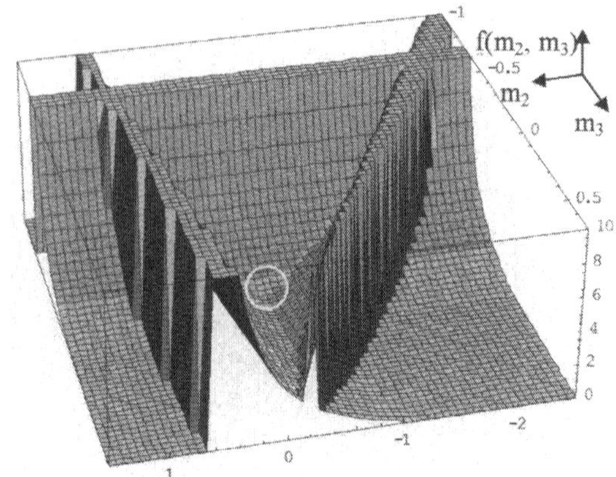

Fig. 4.4. Function f to minimize, for the ThreeWheels rover in a given state. The optimal solution (*circled*), minimizing slip and fulfilling the physical constraints, is found using numerical optimization. The optimal solution corresponds to the intersection of μ_1, μ_2 and μ_3.

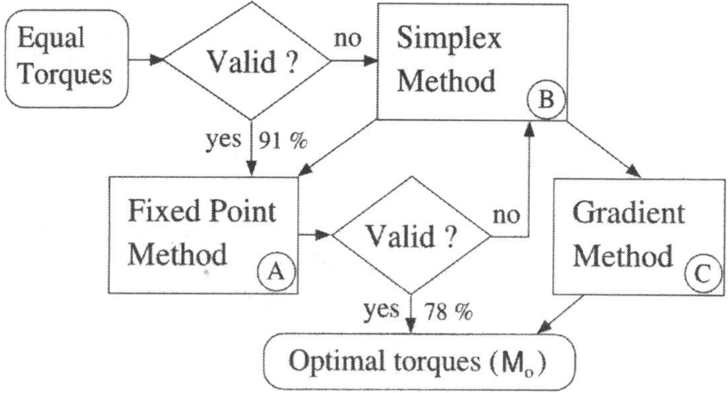

Fig. 4.5. Optimization algorithm. The execution times for the algorithms A, B and C are 6, 5 and 20 ms, respectively (on a 1.5 GHz processor). The worst case is about 31 ms but most cases (71%) are handled by A, which takes only 6 ms.

A. Fixed-point Optimization

This optimization method is based on the fixed-point algorithm. The aim of this algorithm is to numerically find the intersection of functions when an analytical solution is difficult or impossible to obtain. In our case, the optimal solution corresponds to the intersection of μ_1, μ_2 and μ_3. The corresponding flow chart of the algorithm is presented in Fig. 4.6.

This algorithm is computationally light and provides good results for most cases. Nevertheless, it diverges sometimes and does not account for the aforementioned constraints. This can lead to torques that cannot be produced by the motors.

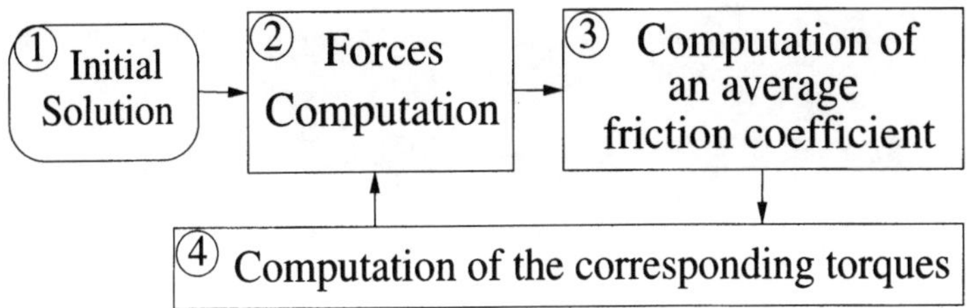

Fig. 4.6. Fixed-point based algorithm. The quasi-static model (**2**) is solved with an initial set of torques (**1**). Then the average of μ_1, μ_2 and μ_3 is computed by (**3**) (the forces provided by (**2**) are used to compute μ_1, μ_2 and μ_3). The torques corresponding to this average are computed in (**4**) and fed back into block (**2**). Twenty iterations are generally sufficient for convergence.

B. Simplex Method

This method is based on the simplex algorithm which solves linear programs in a constrained solution space. The simplex method tries to maximize an objective function considering a set of constraints on the variables. In our case, the algorithm is only used to provide a valid initial solution, and thus, many objective functions can be used. However, in order to get closer to the final optimal solution, we choose the function h defined in 4.15, which tends to minimize the ratio T_i/N_i

$$h = \sum_{i=1}^{n} N_i \tag{4.15}$$

where n is the number of wheels. Furthermore, the function h is linear because it is a linear combination of the torques. The solution provided by this method is guaranteed to fulfill the constraints and can be used as a starting point for both the gradient and fixed-point optimization.

C. Gradient optimization

This algorithm seeks an optimum in the constrained solution space given a known valid initial solution. Gradient optimization is similar to the potential field method: at each step, the gradient is computed and the next solution is generated following the maximum slope.

4.2.3 Torque Optimization for SOLERO

The optimization for the three dimensional SOLERO is identical to the method presented in the previous section. The solution space now has five dimensions and one has to account for 12 constraints

$$N_i > 0 \tag{4.16}$$

$$|m_i| < MaxTrq \tag{4.17}$$

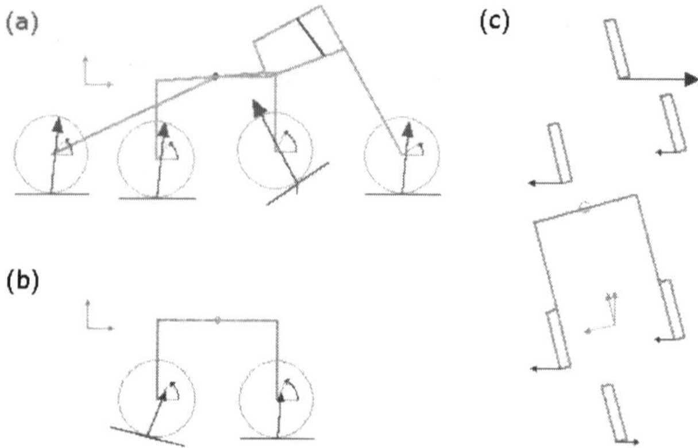

Fig. 4.7. Snapshot of the graphical user interface used to test our optimization procedure. The user can interactively modify the state of the robot, i.e., the wheel-ground contact angles, roll angle, pitch angle and the bogies and fork angles. The optimal torques and forces are computed when the state changes (the forces are expressed in the global reference frame): (**a**) side view; (**b**) right bogie view; (**c**) decomposed view from behind. In this view, the arrows represent the projections of the reaction forces in the global frame of reference.

with $i = 1 \dots 6$ and $MaxTrq$ the saturation torque. An example of computed forces and torques is depicted in Fig. 4.7.

The optimization algorithm has been run for around 20,000 different chassis configurations. The different postures have been generated automatically by varying each parameter to cover the entire robot's configuration space. The parameters are: the wheel-ground contact angles, roll and pitch, the fork, and the angles of the left and right bogies. For each rover state, the minimal friction coefficient (μ_m) has been computed and fed into the histogram shown in Fig. 4.8.

It is interesting to note that, for 80% of the cases, the friction coefficient μ_m is smaller than 0.6 (this corresponds to the static friction coefficient of a tire on a dry road). As explained in Sect. 4.2.1, if μ_m is smaller than the real friction coefficient μ_0, then there is no slip. Thus, there is no slip for 80% of the chassis configurations when the rover is traveling on a terrain with a friction coefficient higher or equal to 0.6 (e.g., a rocky terrain). For more slippery soils it becomes more and more difficult to guarantee no slip. However, the near-exponential shape of the histogram is very favorable and the probability of slip is always minimized whatever the soil type.

The bar of the histogram corresponding to friction coefficients higher than one corresponds to situations where it is impossible to keep static equilibrium and to avoid slip. Indeed, a friction coefficient is always smaller than one by definition. However, this is not critical in practice because, at a higher level, the path planner avoids areas in which the rover risks reaching such extreme chassis configurations [18]. Thus, such cases can be discarded.

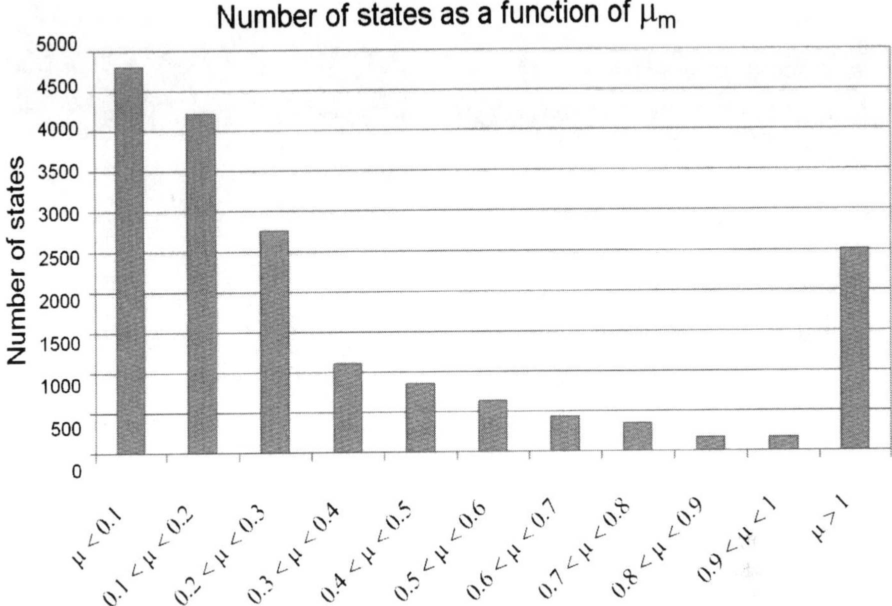

Fig. 4.8. Histogram of μ_m for about 20,000 different robot postures. 80% of the states correspond to a friction coefficient smaller than 0.6 and 65% to a friction coefficient smaller than 0.3.

4.3 Rover Motion

Rolling resistance is another important aspect of the quasi-static model, and is therefore reviewed here. A static model balances the forces and moments of a system to remain at rest or maintain a constant speed. Such a system is an ideal case and does not include resistance to movement. Therefore, an additional torque compensating for the rolling resistance torque must be added to the wheels in order to complete the model and guarantee motion at constant speed. This results in a quasi- static model. Several rolling resistance models are developed in the literature and can be incorporated in the static model to ensure constant speed motion. A rolling resistance model for an elastic wheel on an elastic soil is presented in [37]. Other models applicable for rigid wheels and deformable soils such as sand or earth can be found in [14, 15, 6]. In practice, the parameters of these models are generally difficult to estimate and are valid only for a specific type of soil and condition. Furthermore, the behavior of the controller can be unstable when wrong parameters and/or models are used: what would happen if a controller designed for sand is used on rock? Because an exploration rover is subject to deal with different types of terrain, using a controller endowing a wheel-ground interaction model-specific to one type of soil is generally not appropriate. Figure 4.9 is a good illustration: when driving on such a terrain some wheels might roll on sand and some others on bare rock. Furthermore, the grit and compactness of the sand changes depending on the local conditions.

Fig. 4.9. Images of the Mars surface taken by the NASA rover Spirit next to the Bonneville Crater

In practice, parameterized wheel-soil interaction models cannot be used for such terrain types. Thus, rolling resistance cannot be derived from the models and must be estimated as the robot moves. Our approach introduces a closed-loop controller for estimating rolling resistance and minimizing wheel slip. It is depicted in Fig. 4.10

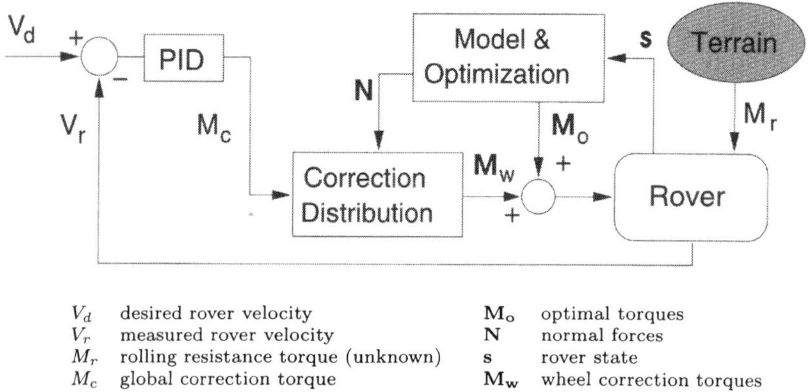

Fig. 4.10. Rover motion control loop. The global loop is a speed-control loop whereas the controllers for the wheels are torque controllers. The rover state vector s includes the wheel-ground contact angles, the internal links angles and the roll and pitch angles.

The kernel of the control loop is a PID controller. It enables the estimation of the additional torques to apply to the wheels in order to reach the rover's desired velocity V_d. The PID minimizes the velocity error $V_d - V_r$, where V_r is the rover's actual velocity measured using onboard sensors[1]. M_c is actually the estimate of

[1] The rover velocity is estimated using the sensor fusion algorithm presented in Chapter 5.

the global rolling resistance torque M_r, which is considered as a perturbation by the PID controller. The rejection of the perturbation is guaranteed by the integral term I of the PID. Because we assume that the rolling resistance is proportional to the normal force, the additional torques to apply to each wheel are calculated using

$$M_{w_i} = \frac{N_i}{N_m} \cdot M_c \tag{4.18}$$

where N_i is the normal force on wheel i and N_m the average of all the normal forces. The derivative term D of the PID allows us to account for nonmodeled dynamic effects and allows stabilization of the system. The same set of parameters can be used for very different terrains because the stability margin offered by such a system is large: the ratio between the rover's inertia and the motor torques is high. Furthermore, we are more interested in minimizing slip than in reaching the desired velocity very precisely. For locomotion in rough terrain, a residual velocity error is accepted as long as slip is minimized.

4.4 Experimental Results

A simulation phase tested the algorithms to verify the theoretical concepts and assumptions. The simulation parameters were set as close as possible to the real operational conditions. However, the intent was not to get exact outputs, but to compare different control strategies.

4.4.1 Simulation Tools

Simulations were realized with the Open Dynamics Engine [4]. This open-source library simulates rigid body dynamics in three dimensions. It has advanced joint types and integrates collision-detection with friction. The source code is available, so it is possible to integrate more sophisticated simulation models such as rolling resistance, friction in the joints, etc. In this application, a rolling resistance proportional to the normal force on the wheel has been implemented.

The simulation tools allow testing and comparison of different traction control strategies. In our experiments, wheel slip is taken as the main benchmark and the performance of our controller (predictive control) is compared to the controller presented in [13] (reactive control). As discussed above, the reactive controller implements speed controllers for the wheels, whereas wheel torques are controlled in the predictive approach.

Wheel Slip Definition

The slip of wheel i at time step k is computed using

$$s_k^i = \Delta w^i_{(k-1,k)} - \Delta\theta^i_{(k-1,k)} \cdot R \tag{4.19}$$

where $\Delta w^i_{(k-1,k)}$ is the true wheel displacement, $\Delta\theta^i_{(k-1,k)}$ the angular change and R the wheel radius. The total slip of the rover integrated during an experiment is defined as

$$S = \sum_k \sum_{i=1}^{6} s^i_k \tag{4.20}$$

Wheel-Ground Contact Angle Estimation

The body collision algorithm of ODE provides n contact points around the wheel together with the forces at those points. These data are similar to those measured using a tactile wheel (the wheel deflection is a function of the applied force) and the same method as presented in Sect. 4.5 is applied to compute the contact angle. In some rare cases, no contact point is provided by ODE at a given time step k. Either the wheel does not touch the ground, or the body collision algorithm fails to compute contact between the wheel and the ground (modeled as a 3D mesh). In this situation, the previous contact angle at time $k-1$ is considered (we assume slow motion and a short simulation period of 10 ms).

4.4.2 Experiments

We set the nominal rover speed for all the simulations to 0.1 m/s and the friction coefficient to 0.6. Different terrain shapes such as depicted in Figs. 4.11 and Fig. 4.12 were generated to perform the experiments in a wide range of conditions. The terrain of Fig. 4.11 is a simple 2D profile (in the plane x-z), whereas

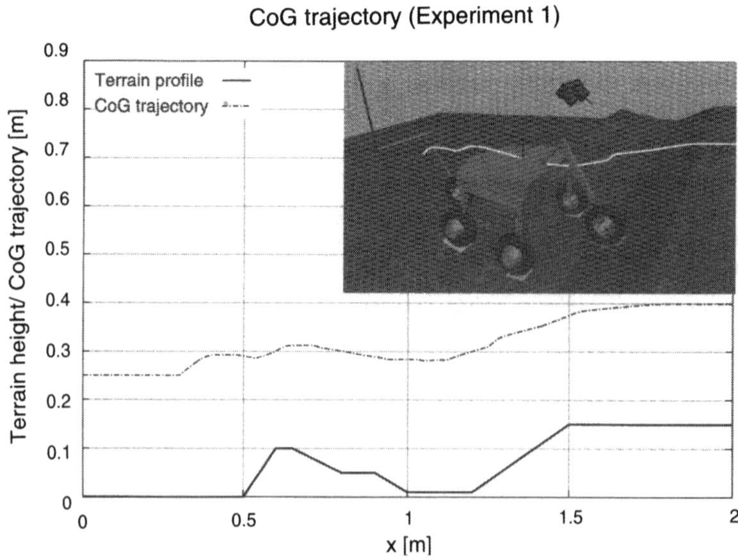

Fig. 4.11. Trajectory of the center of gravity and terrain profile for experiment 1. This kind of terrain is difficult for a wheeled rover because it comprises sharp edges.

Fig. 4.12. Terrain for experiment 2. This kind of terrain is challenging for a wheeled rover because much side slip occurs.

the terrain of Fig. 4.12 is a full 3D mesh. The latter was generated randomly using step, sinusoidal, circle and particle deposition functions.

Because both controllers behave differently in a given situation, the trajectories of the rover can differ significantly. Indeed, the slip of the wheels is not the same for both controllers, and this can cause the rover to take different routes. To allow performance comparison, we consider an experiment as valid if the distance between the final positions of both paths is smaller than a certain threshold (for a path length of about $3.5\,m$). A distance threshold of $0.1\,m$ is small enough to consider the routes as similar.

For all the valid experiments, predictive control showed better performance than reactive control. In some cases, the rover was unable to even climb some obstacles and to reach the final position when driven using the reactive approach. The results for the two examples are presented in Table 4.1. It can be seen that the integrated wheel slip is smaller if predictive control is used.

All the results obtained during the experiments are similar to the one depicted in Figs. 4.13 and 4.14. On flat terrain the performance of both controllers are identical and wheel slip is zero. As soon as bumps and slopes are introduced, the instantaneous slip for each wheel is reduced when using predictive control: the

Table 4.1. Integrated wheel slip S

Experiment	Predictive control	Reactive control	Improvement vs reactive control
1	$S = 0.32\,\mathrm{m}$	$S = 0.48\,\mathrm{m}$	20%
2	$S = 0.61\,\mathrm{m}$	$S = 0.86\,\mathrm{m}$	16%

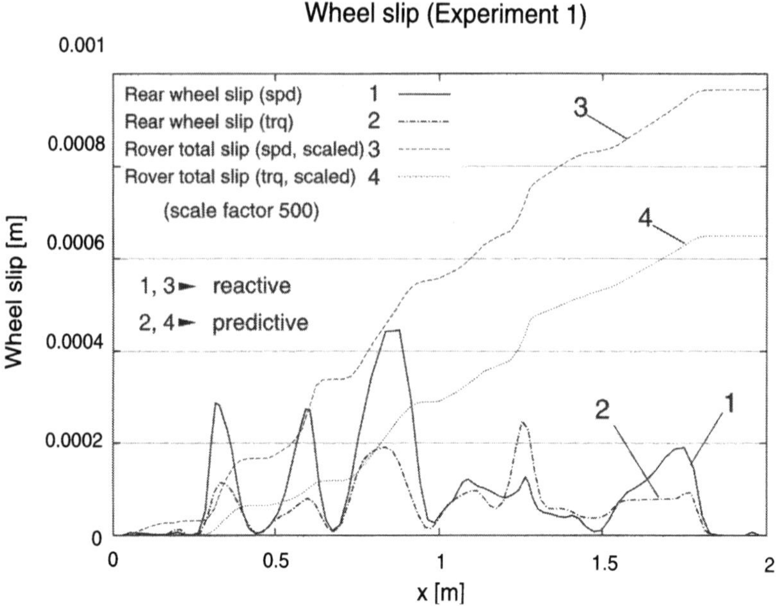

Fig. 4.13. Total slip and rear-wheel slip for experiment 1 (total slip is scaled by a factor of 500). Locally, wheel slip can be bigger with predictive control but the total slip remains always smaller: 20% better than reactive control.

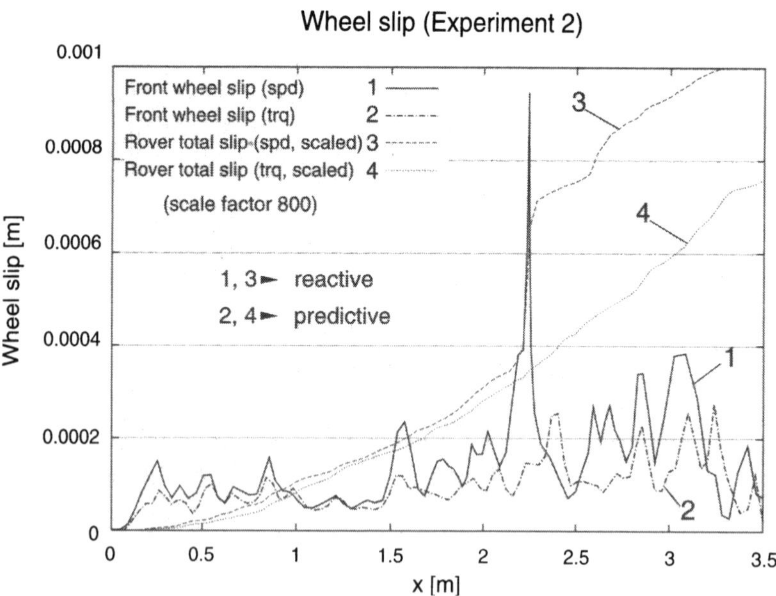

Fig. 4.14. Total slip and front wheel slip for experiment 2 (total slip is scaled by a factor of 800). The difference gets bigger as the rover encounters rough terrain. At the end of the experiment, the predictive controller performs 16% better than the reactive controller.

peaks are at the same places for both controllers, but the amplitude is smaller for the predictive controller. That means that when a wheel slips at a given place, it slips less when predictive control is used. This major behavior can be observed in Figs. 4.13 and 4.14 when looking closely at curves 1 and 2. The peaks may not be aligned but that is mainly due to slight differences in the trajectories. The improvement is difficult to quantify because it strongly depends on the kind of terrain. However, because this behavior has been observed systematically for all the experiments we performed, it validates the benefit of using a predictive controller. As noted above, the focus was on the comparison of the two controllers, not on the absolute figures.

Another important result of the simulations is that they allowed us to test the approach against violations of the strong assumptions that were made during the development of the controller, i.e., the wheels always have ground contact, there is no slip and there is no dynamic effect. During the experiments, these assumptions were all violated but no instability was observed and the controller performed well in all the tested configurations.

Finally, the simulations showed good results and promising perspectives. Furthermore, they allowed us to detect potential problems and address implementation details, bringing us a step closer to the real application.

4.5 Wheel-Ground Contact Angles

The contact angles between the wheels and the ground are key parameters required for traction optimization algorithms. There are different ways of determining these angles. The method presented in Chapter 3 and the one described in [34] are similar because they both consider the displacement and velocity of each wheel for computing the contact angles. The accuracy of the estimation provided by such methods strongly depends on wheel slip and terrain profile. In particular, no estimation can be computed when the rover is at stand-still and poor results are obtained in slowly changing terrain profiles. In [54] the information of the global rover motion is used to estimate the contact angles, limiting the sensitivity to individual wheel slip.

All these approaches assume no slip or account for an accurate velocity estimation. Therefore, they are all subject to the "chicken and egg" problem, i.e., wrong wheel-ground contact angles lead to unadapted motor commands, which cause more wheel slip that generates bad angle estimations. Thus, direct measurement of wheel-ground contact angles is preferred because it is independent of the terrain characteristics and guarantees the system stability. The contact angles can be obtained by measuring the forces acting on the wheels. This can be done using flexible wheels equipped with sensors measuring deflection or by equipping the wheel hub with torque and force sensors. Such approaches have the advantage of providing the contact points for static conditions as well. An example of a tactile wheel is depicted in Fig. 4.15 and more information can be found in [43].

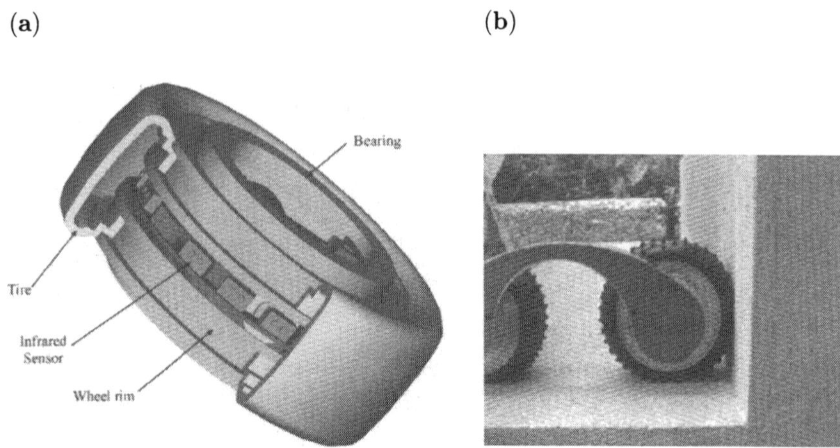

Fig. 4.15. The tactile wheel (developed at EPFL by Michel Lauria). (**a**) Sixteen infrared proximity sensors measure the tire deflection all around the wheel; (**b**) picture of the front wheel of the robot Octopus, equipped with tactile wheels.

With such a wheel, the contact angle is estimated simply using a weighted mean of the proximity sensor signals. In this way a smooth transition of the measured angle is obtained even when sharp slope changes are encountered. In Fig. 4.15b, the force on the tire is transferred from vertical to horizonal as the wheel climbs the step in a continuous manner. The weighted mean translates this behavior in the wheel-ground contact estimation. There are at least two other advantages of including deflection sensors in the wheels:

- improvement of the 3D-Odometry accuracy: a) a direct measurement provides better estimates of the contact angles; b) the sensors measure the effective wheel radii, which are required inputs for 3D-Odometry
- improvement of the controller performance: the deflection of the wheel at a given contact point is an image of the applied force at that point. This information can be incorporated in the model in order to improve the estimates of the normal forces, which are used in 4.18

4.6 Summary

Most of the physics-based control approaches to slip-minimization rely on the knowledge of a specific wheel-soil interaction model. However, in a real application, the model parameters are unknown because the rover is subject to deal with different types of soil such as sand, rocks, gravel or grass. An error in the model parameters has a negative, possibly catastrophic impact on the performance of the controller.

In this chapter, a quasi-static model of a six-wheeled rover together with an optimization method to minimize slip has been presented. Unlike other control strategies, the proposed method does not require the use of soil models, instead it estimates the rolling resistance as the robot moves. As a consequence, the rover is able to operate on different types of soil, which is the main requirement for exploration missions. Furthermore, our approach can be adapted to any kind of wheeled rover and the needed processing power remains low, which enables online computation. The simulations showed that such a controller performs better than a reactive controller and that the system is mature enough to be implemented on the rover for real experiments.

Another interesting aspect of our controller is that the normal forces are explicitly computed and can be used to estimate a slip probability for each wheel: the less pressure on the wheel, the more likely the wheel is to slip. The slip probability can then be propagated through the 3D-Odometry equations to finally obtain the covariance matrix for the robot's displacement. This covariance is valuable information for probabilistic multi-sensor fusion that is presented in the next chapter.

5 Position Tracking in Rough-Terrain

As explained in the introduction, accurate pose tracking is essential for applications where no absolute positioning means is available for extended periods of time. Such conditions hold for underground mining vehicles and planetary exploration rovers, for example. That kind of environment is a challenge for position tracking because of the roughness of the encountered terrain. Indeed, the vibrations, the high-amplitude orientation changes and wheel slip have a negative impact on the quality of sensor measurements and thus on sensor fusion. The previous chapters presented the different actions that have been taken to increase measurement quality. Chapter 2 introduced a platform leading to smooth motion across obstacles. That way, both the vibrations and wheel slip are limited. Then, Chapter 3 and 4 presented two complementary approaches for increasing the accuracy of the odometry. The smoother motion and increased accuracy of the odometry enable sensor fusion in favorable conditions.

The intent of this chapter is to develop a method for combining different sources of information to estimate the six degrees of freedom of a rough-terrain rover. The approach, based on an extended information filter, is presented in Sect. 5.3. Section 5.1 gives a survey of the sensors that can be used in rough terrain and section 5.2 presents the problem of having sensors distributed at different places on the rover. Finally, Sect. 5.4 presents the experimental results, validating the theory.

5.1 Sensor Selection for Motion Perception

The aim of this section is to give a survey of sensors that can be used for position tracking and emphasize the difficulties of motion perception in unstructured outdoor environments.

The family of 1D/2D distance sensors such as ultrasound and 2D laser scanners are commonly used indoors (structured environments), but are not well adapted for outdoors (unstructured environments). In structured environments, it is generally assumed that a rover moves on flat ground and that its working space is delimited by walls, perpendicular to the ground. This strong assumption

P. Lamon: 3D-Position Track. & Cntrl. for All-Terrain Robots, STAR 43, pp. 53–79, 2008.
springerlink.com

allows using a simple world representation, and distance information is used to localize the rover. Simple features such as corners, lines and segments are extracted from raw data and are relevant enough for localization [66, 8]. When dealing with unstructured environments (3D world), these sensors do not provide enough information and the interpretation of their measurements is tedious because of the lack of *a priori* knowledge.

Monocular vision (a single camera) provides a lot of information and covers a large field-of-view: from very close distances to the horizon (this is not the case for distance sensors, which are limited to a predefined range). Many applications use monocular vision as a source of information for position tracking. When combined with a panoramic mirror, e.g., parabolic or equiangular, the field-of-view of a camera is extended to 360°. Such a panoramic vision system is used in [62] for tracking the six degrees of freedom of a robotic platform. In [23], the skylines extracted from panoramic views are used to localize a mobile robot, provided a topographic map is available. However, monocular vision provides only scaleless motion information. This information has to be completed with metric data to obtain metric estimations.

Range imagers such as stereovision are valuable sensors for outdoor applications. For example, the MER rovers are equipped with two stereovision systems that are used for terrain traversability analysis and visual odometry [27, 22]. The principle of visual odometry is to compute an estimate of the robot's motion on the base of a set of 3D point-to-point matches (see Appendix D). However, the use of stereovision has some limitations: it works well only in textured environments, its range is limited and the images can be affected by bad illumination conditions, leading to poor visual odometry estimations.

Odometry based on wheel encoders is a well-known technique for motion estimation. In Chapter 3, it has been shown that odometry can be used on uneven terrains if the mechanical structure of the rover is adapted. However, because of wheel slip, the position estimation error can grow quickly and this technique cannot be used as the only means to track a rover's position.

The inertial sensors measure the accelerations and rotation rates of a body on which they are mounted. Thus, the 3D position and the orientation of a rover can be tracked by performing the double integration of the accelerations and the single integration of the angular rates. In the presence of a known gravity field, the attitude angles of the rover, i.e., roll and pitch, can be estimated in an absolute manner. However, the pose estimation and the heading diverge quickly because the signals are affected by biases and scaling errors that are integrated over time. Thus, such sensors cannot be used alone to estimate motion and have to be combined with other sensors to limit drift.

Heading sensors such as compasses are of high interest because they provide absolute heading information and therefore are not subject to drift. However, the magnetic field is generally not homogeneous, or even usable on the surface of Mars. Sun sensors and star sensors use the sun and the stars, respectively, as absolute references (the design of a sun sensor is presented in [67]). Both provide absolute heading, and the star sensors also provide latitude and longitude. These sensors can only be used when the rover is perfectly still, and require good meteorological conditions. Furthermore, a sun sensor can be utilized only during the day, whereas the measurements of a star sensor are only available during the night. However, they are of high interest for global re-localization: the absolute position acquired by a star sensor during the night can be used to relocate the rover globally after a long traverse. Thus, three conclusions can be drawn from the above discussion:

1. No single sensor can provide all the required information. All have their own drawbacks and advantages. In general, the value of the provided information is inversely proportional to the sampling rate, e.g., an inertial measurement unit can provide data at 100 Hz but the heading estimation diverges quickly, whereas a sun sensor provides absolute heading but requires the rover to remain at rest over a longer measurement time.
2. Since the data provided by absolute sensors contain no drift, they have more value than those acquired using dead reckoning sensors.
3. Because a small angle error (e.g., heading error) leads to a large position error, it is more important to have precise information about angles than about distances.

These conclusions enforce the fact that the use of complementary sensors is required for robust position tracking. In this chapter, we use three different sources of information for sensor fusion, though our approach can be easily extended to more sources:

- Wheel encoders: as discussed in Chapter 3, odometry can be used to predict the robot's displacement and a reasonably accurate motion estimation can be obtained in rough terrain. Moreover, Chapter 4 proposes an algorithm to limit wheel slip and enhance the accuracy of odometry. This motion estimation technique provides data at a relatively high rate (10 Hz).
- IMU: an inertial measurement unit enables the direct measurement of system dynamics. When a gravity field is present, the attitude can be estimated without any drift, which is very valuable information. Furthermore, the measurements are available at a high frequency (75 Hz).
- Stereovision: a method similar to [48, 53] is used to estimate the six degrees of freedom of the rover. Instead of using pixel tracking, interest points are extracted for each frame and are matched using the algorithm presented in [35, 36]. This technique, called Visual Motion Estimation (VME), usually provides better pose estimation than the other sensors but at a much lower rate (0.5 Hz).

For this research, absolute sensors such as GPS have not been considered because the rover is supposed to track its position in an unknown environment without the help of any artificial beacons (e.g., exploration of Mars).

5.2 Uncertainties Propagation

A mobile robot is equipped with several sensors positioned at different places on its chassis. In order to enable sensor fusion, a common reference frame must be chosen. Then, the position of each sensor has to be expressed in this same coordinate system. The following sections develop the equations for propagating the uncertainties associated with the sensors' measurements (incremental motion information) through coordinate system transformations.

5.2.1 Coordinate Systems and Transformations

The position of a sensor S is expressed by a homogeneous transformation matrix

$$H = \begin{bmatrix} c\theta c\psi & -c\phi s\psi + s\phi s\theta c\psi & s\phi s\psi + c\phi s\theta c\psi & x_t \\ c\theta s\psi & c\phi c\psi + s\phi s\theta s\psi & -s\phi c\psi + c\phi s\theta s\psi & y_t \\ -s\theta & s\phi c\theta & c\phi c\theta & z_t \\ 0 & 0 & 0 & 1 \end{bmatrix} \tag{5.1}$$

where s and c denote, respectively, the sin(.) and cos(.) functions. Such a matrix includes both rotation and translation in three dimensions. The rotation is a three by three direction cosine matrix expressed in terms of the angles ϕ, θ, ψ and the components of the translation are x_t, y_t and z_t. Thus, such a transformation is parametrized by six parameters that are the six degrees of freedom of a body in a 3D space. It is interesting to note that a homogeneous matrix can also be used to describe incremental motion (in a time interval t, $t+1$) or even to describe the robot's pose as illustrated in Fig. 5.1. Between times t and $t+1$, the sensor moves from point S to S'. The 3D transformation reflecting the sensor motion is given by $H_{S'S}$. Thus, the six parameters defining $H_{S'S}$ include all the motion information between t and $t+1$ (expressed in the sensor frame). We define $\mathbf{p_S}$, the vector of the parameters defining $H_{S'S}$

$$\mathbf{p_S} = \begin{bmatrix} \phi_S \; \theta_S \; \psi_S \; x_S \; y_S \; z_S \end{bmatrix}^T \tag{5.2}$$

As explained at the beginning of this section, we need to express this transformation in a common coordinate system, which in our case is the coordinate system linked to the robot's body ($Rx_ry_rz_r$). In other words, we need to compute the motion of the robot's center R, i.e. $H_{R'R}$. Considering that the position of the sensor relative to the robot's frame remains constant ($H_{SR} = H_{S'R'}$), one can write

$$x_R = H_{SR}\, H_{S'S}\, x_s = H_{R'R}\, H_{SR}\, x_s \tag{5.3}$$

where x_s is a point in the sensor frame at time $t+1$ and x_R its coordinates in the robot's frame at time t. Using 5.3, the motion of the robot expressed in the robot's coordinate system is given by

$$H_{R'R} = H_{SR}\, H_{S'S}\, H_{SR}^{-1} \tag{5.4}$$

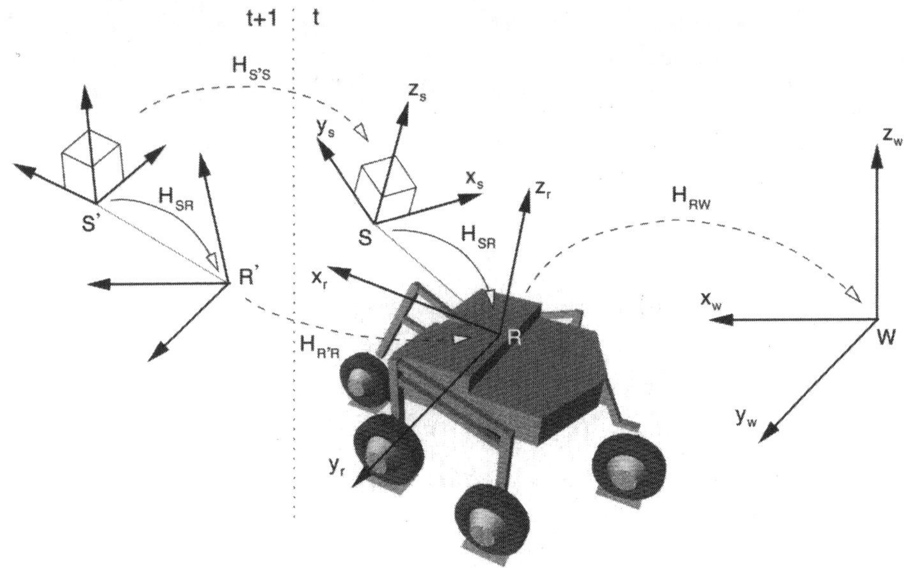

Fig. 5.1. Transformations between the different coordinate systems. W, R and S are, respectively, the centers of the coordinate systems associated with the *World*, the *Robot* and the *Sensor*. The homogeneous matrix H_{ij} allows the transformation of coordinates expressed in the reference frame i into that of frame j. "Prime" signs added to variable names (e.g., p') denote values related to time-step $t + 1$.

and the corresponding parameters vector is defined as

$$\mathbf{p_R} = \begin{bmatrix} \phi_R \; \theta_R \; \psi_R \; x_R \; y_R \; z_R \end{bmatrix}^T \tag{5.5}$$

Using a similar approach, the pose of the robot at time $t+1$, expressed in the world coordinates system, is given by

$$H_{R'W} = H_{RW} \; H_{R'R} \tag{5.6}$$

Finally, we define $\mathbf{p_W}$ and $\mathbf{p'_W}$ as the parameters vectors of H_{RW} and $H_{R'W}$

$$\mathbf{p_W} = \begin{bmatrix} \phi_W \; \theta_W \; \psi_W \; x_W \; y_W \; z_W \end{bmatrix}^T \tag{5.7}$$

$$\mathbf{p_{W'}} = \begin{bmatrix} \phi'_W \; \theta'_W \; \psi'_W \; x'_W \; y'_W \; z'_W \end{bmatrix}^T \tag{5.8}$$

5.2.2 Error Propagation

For sensor fusion, we need to assess how the uncertainties associated with the sensor measurement $\mathbf{p_S}$ propagate[1] through the coordinate system transformation

[1] An introduction to error propagation is given in [8, 49].

$H_{R'R}$. The uncertainties associated with the transformation H_{SR} are neglected because H_{SR} can be calibrated accurately. In the latter, we define C_S and C_R as the covariance matrices associated with the vectors $\mathbf{p_S}$ and $\mathbf{p_R}$ respectively. In order to propagate these covariances, the function q, expressing $\mathbf{p_R}$ as a function of $\mathbf{p_S}$ has to be derived. q is determined using 5.4 and the properties of the homogeneous transformation matrices

$$\begin{aligned} q_0(\mathbf{p_S}) &= \arctan(H_{R'R}(3,2)/H_{R'R}(3,3)) = \phi_R \\ q_1(\mathbf{p_S}) &= \arcsin(-H_{R'R}(3,1)) = \theta_R \\ q_2(\mathbf{p_S}) &= \arctan(H_{R'R}(2,1)/H_{R'R}(1,1)) = \psi_R \\ q_3(\mathbf{p_S}) &= H_{R'R}(4,1) = x_R \\ q_4(\mathbf{p_S}) &= H_{R'R}(4,2) = y_R \\ q_5(\mathbf{p_S}) &= H_{R'R}(4,3) = z_R \end{aligned} \tag{5.9}$$

Then the covariance matrix associated with $\mathbf{p_R}$ is given by

$$C_R = J_S\, C_S\, J_S^T \qquad \text{with} \quad J_S = \frac{\partial q}{\partial \mathbf{p_S}} \tag{5.10}$$

Now, we are interested in computing the uncertainty associated with the robot's pose at time $t+1$. For that, we need to combine the uncertainty of the pose at time t and the uncertainty associated with the incremental motion $\mathbf{p_R}$. The function q' (a set of six functions), expressing $\mathbf{p_{W'}}$ as a function of $\mathbf{p_W}$ and $\mathbf{p_R}$ is obtained using 5.6 and the properties of the homogeneous transformation matrices

$$\begin{aligned} q'_0(\mathbf{p_W},\mathbf{p_R}) &= \arctan(H_{R'W}(3,2)/H_{R'W}(3,3)) = \phi_{W'} \\ q'_1(\mathbf{p_W},\mathbf{p_R}) &= \arcsin(-H_{R'W}(3,1)) = \theta_{W'} \\ q'_2(\mathbf{p_W},\mathbf{p_R}) &= \arctan(H_{R'W}(2,1)/H_{R'W}(1,1)) = \psi_{W'} \\ q'_3(\mathbf{p_W},\mathbf{p_R}) &= H_{R'W}(4,1) = x_{W'} \\ q'_4(\mathbf{p_W},\mathbf{p_R}) &= H_{R'W}(4,2) = y_{W'} \\ q'_5(\mathbf{p_W},\mathbf{p_R}) &= H_{R'W}(4,3) = z_{W'} \end{aligned} \tag{5.11}$$

Finally, the covariance matrix of the pose at time $t+1$ is given by

$$C_W^{t+1} = J_R\, C_R\, J_R^T + J_W\, C_W^t\, J_W^T \tag{5.12}$$

where C_W^t is the covariance matrix associated with $\mathbf{p_W}$ and

$$J_R = \frac{\partial q'}{\partial \mathbf{p_R}} \qquad\qquad J_W = \frac{\partial q'}{\partial \mathbf{p_W}} \tag{5.13}$$

5.3 Sensor Fusion

Probabilistic data fusion is the most-used method for combining uncertain information. All the probabilistic filters such as the Hidden Markov Models, the

Partially Observable Markov Decision Processes, or the Kalman filter are derived from the Bayes rule and provide a framework to fuse uncertain data [10]. The choice of one or another depends on the application. For continuous state spaces, the Kalman filter is the preferred choice for sensor fusion. Even if this method can be applied to fuse the measurements acquired by any number of sensors, most of the applications found in the literature generally use only two sensors. The most commonly used pairs are: odometry/laser scanner, odometry/inertial measurement unit [19], inertial measurement unit/vision [63, 56, 71], compass/inertial measurement unit [55], inertial/GPS [51], etc. Furthermore, even for rough terrain rovers, only the 2D case (x, y, ψ) is generally considered.

In this section, a method to estimate the six degrees of freedom of the rover (i.e. $x, y, z, \phi, \theta, \psi$) using the measurements of three different sensors is presented. Its principle is presented in Fig. 5.2. Our approach uses an extended information filter (EIF) to combine the information acquired by the sensors. This formulation of the Kalman filter has very interesting features: its mathematical expression is well suited to implement a distributed sensor fusion scheme and allows for easy extension of the system in order to accommodate any number of sensors, of any kind [49]. In this application, the observation and transition equations are not linear and a nonlinear information filter is required. The observation equation assumes additive zero mean Gaussian noise and is written

$$z_j(k) = h_j\left[k, x(k)\right] + \nu_j(k) \tag{5.14}$$

where z_j is the measurement vector of sensor j and h_j the non-linear observation model transforming the state vector $x(k)$ from the state space to the observation space. We define the measurement covariance matrix as being the expectation of the measurement noise: $R_j = E\left\{\nu_j\, \nu_j^T\right\}$. Similarly, the nonlinear state transition equation can be written as

$$x(k) = f\left[k, x(k-1)\right] + \omega(k) \tag{5.15}$$

where f is the non-linear state transition model describing the transition of the state from one time-step to another as a nonlinear function of the state. The covariance matrix of the state transition is defined as $Q = E\left\{w\, w^T\right\}$. The first order nonlinear information filter is similar to the linear information filter if the following substitutions are made

$$\nabla_x h_j\left[k, \hat{x}(k)\right] \equiv H_j(k) \tag{5.16}$$

$$\nabla_x f\left[k, \hat{x}(k)\right] \equiv F(k) \tag{5.17}$$

For the information filter, the information state vector y and the information matrix, which is the inverse of the covariance matrix P, are updated according to

$$\hat{y}(k|k) = \hat{y}(k|k-1) + \sum_{j \in S} H_j^T(k) R_j^{-1}(k) z_j'(k) = P^{-1}(k|k)\, \hat{x}(k|k) \tag{5.18}$$

$$P^{-1}(k|k) = P^{-1}(k|k-1) + \sum_{j \in S} H_j^T(k) R_j^{-1} H_j(k) \tag{5.19}$$

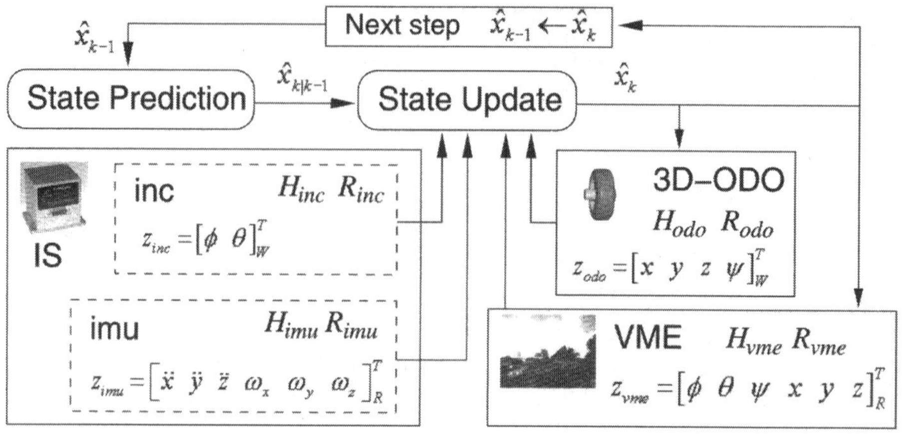

Fig. 5.2. Sensor fusion scheme. The Inertial Sensor (IS) is divided into two logical sensors: an inclinometer (inc) and an inertial measurement unit (imu).

with $\mathrm{S} = \{\mathrm{imu}, \mathrm{inc}, \mathrm{odo}, \mathrm{vme}\}$ as the set of sensors and $\mathrm{z}'(\mathrm{k}) = \mathrm{z}(\mathrm{k})$ for the linear filter, otherwise

$$\mathrm{z'}_\mathrm{j}(\mathrm{k}) = z_\mathrm{j}(\mathrm{k}) - (\mathrm{h_j}\,[k, \hat{x}(k|k-1)] - \nabla_x h_j\,[k, \hat{x}(k|k-1)]\,\hat{x}(k|k-1)) \quad (5.20)$$

The covariance matrix and the information vector are predicted as

$$P(k|k-1) = F(k)P(k-1|k-1)F^T(k) + Q(k) \quad (5.21)$$

$$\hat{y}(k|k-1) = P^{-1}(k|k-1)F(k)P(k-1|k-1)\hat{y}(k-1|k-1) \quad (5.22)$$

Finally, the state vector may be obtained from

$$\hat{x}(k|k) = P(k|k)\,\hat{y}(k|k) \quad (5.23)$$

It is interesting to note that a sensor j can be easily incorporated in the sensor fusion process if the observation model h_j, the covariance R_j and the measurement vector z_j are provided. With the information filter, the update stage is additive. The update of the information vector and the information matrix takes the form of the simple equations 5.18 and 5.19 that can be interpreted this way: the information at time k is equal to the information at time $k-1$ plus the total amount of the information provided by the sensors.

5.3.1 Sensor Models

To implement such a filter, the relationship between the measurement vectors and the state vector has to be determined for all the sensors. The measurement models h_j together with their linear matrix form H_j are presented in this section whereas the methodology for setting up the covariance matrices R_j is discussed in the experimental results section. Indeed, the values in R_j are specific to the sensors used in the experiments. The measurement vectors and matrices for all the sensors are defined in Fig. 5.2.

Inertial Measurement Unit Model

The position, velocity and attitude can be computed by integrating the measurement acquired by the IMU, that is

$$z_{imu} = \begin{bmatrix} \ddot{x} \; \ddot{y} \; \ddot{z} \; \omega_x \; \omega_y \; \omega_z \end{bmatrix} \tag{5.24}$$

However, the accelerometers and gyroscopes are influenced by bias errors. Even if these offsets are small they will cause an unbounded growth in the error of integrated measurements. The velocity and the angles error grow proportionally over time and the position error to the square of time. To limit the drift, these biases are introduced in the state vector for being estimated by the filter. The accelerometers are thus modeled as

$$\begin{bmatrix} \ddot{x} \\ \ddot{y} \\ \ddot{z} \end{bmatrix}_R^z = \Gamma_{WR} \begin{bmatrix} \ddot{x} \\ \ddot{y} \\ \ddot{z} \end{bmatrix}_W + \begin{bmatrix} b_{ax} \\ b_{ay} \\ b_{az} \end{bmatrix}_R + v_a \tag{5.25}$$

and the gyroscopes as

$$\begin{bmatrix} \omega_x \\ \omega_y \end{bmatrix}^z = \begin{bmatrix} \omega_x \\ \omega_y \end{bmatrix} + \begin{bmatrix} b_{\omega x} \\ b_{\omega y} \end{bmatrix} + v_\omega \tag{5.26}$$

$$\omega_z^z = (1 + \Delta_{\omega z}) \left[\omega_z + b_{\omega z}\right] + v_{\omega z} \tag{5.27}$$

Γ_{WR} is the direct cosine rotation matrix that transforms values expressed in the world-fixed coordinate system W into the robot's coordinate system R. This matrix is a function of the angles ϕ (roll), θ (pitch) and ψ (yaw). The b's and v's are the biases and the measurement noises of the signals, respectively. Unlike roll and pitch, the heading of the rover is not periodically updated with absolute data. Therefore, in order to limit error propagation, a special provision is included in the z-gyroscope model: a more accurate modeling, incorporating the scaling error $\Delta_{\omega z}$. Equations 5.25 and 5.26 are nonlinear and the first order Taylor expansion is used to obtain the linearized equations

$$\begin{bmatrix} \ddot{x} \\ \ddot{y} \\ \ddot{z} \end{bmatrix}_R^z = \bar{\Gamma}_{WR} \begin{bmatrix} \ddot{x} \\ \ddot{y} \\ \ddot{z} \end{bmatrix}_W + \nabla \left[\bar{\Gamma}_{WR}\, \bar{a}\right] \begin{bmatrix} \phi \\ \theta \\ \psi \end{bmatrix} + \begin{bmatrix} b_{ax} \\ b_{ay} \\ b_{az} \end{bmatrix}_R + \nu_a \tag{5.28}$$

$$\text{with } \bar{a} = \begin{bmatrix} \bar{\ddot{x}} \; \bar{\ddot{y}} \; \bar{\ddot{z}} - g \end{bmatrix}^T$$

$$\omega_z^z = (1 + \bar{\Delta}_{\omega z})\,\omega_z + (1 + \bar{\Delta}_{\omega z})\,b_{\omega z} + (\bar{\omega}_z + \bar{b}_{\omega z})\,\Delta_{\omega z} + \nu_{\omega z} \tag{5.29}$$

where the bars denote operating point values and g is the gravitational constant, which has to be removed before integrating the accelerations. Then, the matrix H_{imu} can be constructed using 5.26, 5.28 and 5.29 (the detailed linearized equations for the IMU are developed in Appendix B). H_{inc} is the identity matrix because the inclinometer directly measures ϕ and θ (the attitude of the robot).

Because the IMU is not placed exactly at the center of the robot, it is subject to centripetal accelerations due to the angular velocities. These perturbations have to be subtracted from the measurements in order to consider the accelerations related to the center of the robot, which is used as the reference point by all sensors. The centripetal perturbation c_i for each accelerometer is given by

$$c_i = \boldsymbol{\omega} \times (\boldsymbol{\omega} \times \mathbf{r_i}) + \dot{\boldsymbol{\omega}} \times \mathbf{r_i} \tag{5.30}$$

$$\text{with } \boldsymbol{\omega} = \begin{bmatrix} \omega_x \; \omega_y \; \omega_z \end{bmatrix}^T \text{ and } i \in \{x, y, z\}$$

where $\mathbf{r_i}$ is the position of each accelerometer i with respect to the robot's center.

3D-Odometry Measurement Model

The robot used for this research is a partially skid-steered rover and the natural and controlled motion is mainly in the forward direction. Thus, the motion estimation errors due to wheel slip and wheel diameter variations have much more effect in the x-z plane of the rover than along the lateral direction y. Therefore, scaling errors (Δ_{ox} and Δ_{oz}) modeling wheel slip and wheel diameter change are introduced only for the x and z-axes.

3D-Odometry provides an incremental measurement of the rover's motion between time t and $t+1$: $\mathbf{p}_{odo} \begin{bmatrix} d_{ox} \; d_{oy} \; d_{oz} \; d_{o\psi} \end{bmatrix}^T$ (expressed in the robot's coordinate system). Thus, the corresponding transformation matrix $H_{R'R}$ (see Fig. 5.1) is obtained by making the following substitution in 5.1

$$\begin{aligned} x_t &= (1 + \Delta_{ox})d_{ox} \qquad & y_t &= d_{oy} \qquad & z_t &= (1 + \Delta_{oz})d_{oz} \\ \phi &= 0 & \theta &= 0 & \psi &= d_{o\psi} \end{aligned} \tag{5.31}$$

We set the roll and pitch increments to zero because the information about these angles is not explicitly provided by odometry. As the odometry is updated at a relatively high rate, we can consider the small angles approximation. Thus, setting these angles to zero has minimal impact on the incremental motion error.

The position in the world coordinate system is computed with 5.6, using the robot pose at time t and the incremental motion $H_{R'R}$. Finally, 5.11 is used to find the relations between the state vector and the measurement vector. These expressions are not linear and the Jacobian has to be developed to finally obtain H_{odo}.

VME Measurement Model

VME estimates the incremental camera motion between two stereo pairs acquisitions, i.e., $\mathbf{p}_{vme} = \begin{bmatrix} d_{vx} \; d_{vy} \; d_{vz} \; d_{v\phi} \;\; d_{v\theta} \;\; d_{v\psi} \end{bmatrix}^T$. This transformation, expressed in the camera coordinate system, is first converted into the robot's coordinate system using 5.4. Then the same method as presented in the previous section is applied to derive H_{vme}.

5.3.2 State Prediction Model

The state prediction model determines the transition of the state vector from one time-step to another. In our case, it has the following form

$$\begin{bmatrix} \mathbf{x}_x \\ \mathbf{x}_y \\ \mathbf{x}_z \\ \mathbf{x}_{ba} \\ \mathbf{x}_\omega \\ \mathbf{x}_{b\omega} \\ \mathbf{x}_\Delta \end{bmatrix}_{k+1} = \begin{bmatrix} F_x & & & & & & \\ & F_y & & \cdots & & 0 & \\ & & F_z & & & & \\ & \vdots & & F_{ba} & & \vdots & \\ & & & & F_\omega & & \\ & 0 & & \cdots & & F_{b\omega} & \\ & & & & & & F_\Delta \end{bmatrix} \begin{bmatrix} \mathbf{x}_x \\ \mathbf{x}_y \\ \mathbf{x}_z \\ \mathbf{x}_{ba} \\ \mathbf{x}_\omega \\ \mathbf{x}_{b\omega} \\ \mathbf{x}_\Delta \end{bmatrix}_k + \begin{bmatrix} w_x \\ w_y \\ w_z \\ w_{ba} \\ w_\omega \\ w_{b\omega} \\ w_\Delta \end{bmatrix}_k \tag{5.32}$$

with

$$\begin{aligned} \mathbf{x} &= [\mathbf{x}_x\ \mathbf{x}_y\ \mathbf{x}_z\ \mathbf{x}_{ba}\ \mathbf{x}_\omega\ \mathbf{x}_{b\omega}\ \mathbf{x}_\Delta]^T \\ &= [\ddot{x}\,\dot{x}\,x\ \ \ddot{y}\,\dot{y}\,y\ \ \ddot{z}\,\dot{z}\,z\ \ b_{ax}\,b_{ay}\,b_{az}\ \ \omega_x\,\phi\,\omega_y\,\theta\,\omega_z\,\psi\ \ b_{\omega x}\,b_{\omega y}\,b_{\omega z}\ \ \Delta_{\omega z}\,\Delta_{ox}\,\Delta_{oz}]^T \end{aligned}$$

The angular rates, biases, scaling errors and accelerations are random processes that are affected by the motion commands of the rover, time and other unmodeled parameters. However, they cannot be considered as pure white noise exclusively because they are highly time-correlated. In order to illustrate this statement, let us assume that the rover is subject to an acceleration of $2g$ at time t. At time $t+1$, the acceleration cannot reach $-2g$ because the rover has a certain inertia and the elapsed time between t and $t+1$ is short. Thus, these signals are time-correlated and cannot be considered as white noise. Instead, they can be modeled as first order Gauss–Markov processes[2] whose auto-correlation function is

$$R_p(t) = \sigma^2 e^{-\tau|t|} \tag{5.33}$$

where $1/\tau$ is the time constant defining the correlation time and σ^2 is the variance of the process. Such a process can also be considered as a low pass filter, with τ

[2] The detailed derivation of the equations related to the first and second integral of a Gauss–Markov process is presented in Appendix C. In particular, the covariance matrix associated with such a process is developed.

being the time constant. The discrete differential equations of the first and second integral of such a process are computed using the inverse Laplace operator

$$\begin{bmatrix} p_1 \\ p_2 \\ p_3 \end{bmatrix}_{k+1} = \begin{bmatrix} e^{-\tau h} & 0 & 0 \\ (1-e^{-\tau h})/\tau & 1 & 0 \\ (\tau h - 1 + e^{-\tau h})/\tau^2 & h & 1 \end{bmatrix} \begin{bmatrix} p_1 \\ p_2 \\ p_3 \end{bmatrix}_k = \Phi(\tau, h) \begin{bmatrix} p_1 \\ p_2 \\ p_3 \end{bmatrix}_k \tag{5.34}$$

where p_2 and p_3 are, respectively, the first and second integral of the Gauss–Markov process p_1, and h is the sampling time. It is interesting to note that if τ tends toward zero, and if p_1 corresponds to an acceleration, then 5.34 is written as

$$\begin{bmatrix} \ddot{x} \\ \dot{x} \\ x \end{bmatrix}_{k+1} = \begin{bmatrix} 1 & 0 & 0 \\ h & 1 & 0 \\ h^2/2 & h & 1 \end{bmatrix} \begin{bmatrix} \ddot{x} \\ \dot{x} \\ x \end{bmatrix}_k \tag{5.35}$$

This corresponds to the well-known set of equations that represents uniformly-accelerated motion. Thus, the state propagation along x between k and $k+1$ is nothing more than an accelerated motion using the best estimate of the acceleration at time step k. The transition matrices for $\mathbf{x}_x, \mathbf{y}_y$ and $\mathbf{z}_z$ can thus be written as

$$F_x = \Phi(\tau_x, h) \qquad F_y = \Phi(\tau_y, h) \qquad F_z = \Phi(\tau_z, h) \tag{5.36}$$

The biases and scaling factors are simply propagated through a low-pass filter using

$$\begin{aligned} F_{ba} &= \mathrm{diag}(e^{-\tau_{bax}}, e^{-\tau_{bay}}, e^{-\tau_{baz}}) \\ F_{b\omega} &= \mathrm{diag}(e^{-\tau_{b\omega x}}, e^{-\tau_{b\omega y}}, e^{-\tau_{b\omega z}}) \\ F_{\Delta} &= \mathrm{diag}(e^{-\tau_{\Delta\omega z}}, e^{-\tau_{\Delta ox}}, e^{-\tau_{\Delta oz}}) \end{aligned} \tag{5.37}$$

where $\mathrm{diag}(a, b, c)$ refers to a diagonal matrix composed of the elements a, b and c. The covariance matrix Q_p associated with a Gauss–Markov process and its integral terms are derived by computing the individual expectations $E\{p_i\, p_j\}$ with $i, j = 1 \ldots 3$. Thus, because the accelerations, biases and scaling errors are modeled as Gauss–Markov processes, one can write

$$Q_x = E\left\{w_x w_x^T\right\} \qquad Q_y = E\left\{w_y w_y^T\right\} \qquad Q_z = E\left\{w_z w_z^T\right\} \tag{5.38}$$

$$Q_{ba} = E\left\{w_{ba} w_{ba}^T\right\} \qquad Q_{b\omega} = E\left\{w_{b\omega} w_{b\omega}^T\right\} \qquad Q_{\Delta} = E\left\{w_{\Delta} w_{\Delta}^T\right\} \tag{5.39}$$

The derivation of F_ω is more tedious because the dynamics of x_ω is nonlinear. Furthermore, the small-angle approximation cannot be made because the robot moves on rough terrain, where angular variations can be of high amplitude. Equation 5.40 describes the nonlinear state transition of x_ω

$$
\begin{aligned}
&\omega_{x_{k+1}} = \omega_{x_k} e^{-\tau_{\omega_x} h} \quad \omega_{y_{k+1}} = \omega_{y_k} e^{-\tau_{\omega_y} h} \quad \omega_{z_{k+1}} = \omega_{z_k} e^{-\tau_{\omega_z} h} \\
&\phi_{k+1} = q_1(x_\omega) = \phi_k + h((\omega_y \sin\phi + \omega_z \cos\phi)\tan\theta + \omega_x) \\
&\theta_{k+1} = q_2(x_\omega) = \theta_k + h(\omega_y \cos\phi - \omega_z \sin\phi) \\
&\psi_{k+1} = q_3(x_\omega) = \psi_k + h(\omega_y \sin\phi + \omega_z \cos\phi)/\cos\theta
\end{aligned}
\tag{5.40}
$$

Finally, the linearized 6x6 matrix F_ω is obtained by computing the Jacobian of the q functions (see Appendix B.2).

5.4 Experimental Results

The aim of this section is to describe the methodology used to define the state transition matrix Q and the measurement covariance matrices R_j, and to validate the theory through experiments conducted on SOLERO. In order to better illustrate how each sensor contributes to the pose estimation and in which situation, the experiments were divided into two parts. The first part describes the results of sensor fusion between inertial sensors and odometry only, whereas the second part involves all three sensors, i.e., odometry, inertial sensor and visual motion estimation.

5.4.1 Inertial Sensor and 3D-Odometry

An IMU provides direct measurements of the dynamics of a system and is self-contained. For these reasons, it is used in many applications to estimate the robot's velocities and orientation. IMUs were first used in aerospace and a large part of the literature refers to such applications, (see [65] for theory and applications). The availability of integrated low-cost and low-power solid-state sensors enabled the usage of IMUs for ground applications such as mobile robotics. Nevertheless, their implementation on ground vehicles is more difficult than on aircraft because of the poor signal-to-noise ratio; the vehicles move relatively slowly and are subject to vibrations.

Many research works are related to road vehicle applications, in which an IMU is used to provide a higher update rate of the position between two consecutive GPS data acquisitions. In particular, this combination of sensors is used in [73, 17] to estimate the wheel diameter changes and the vehicle sideslip. Reference [12] shows that a low-cost IMU can improve the localization system performance and can be applied to mobile robotics if an accurate sensor model is provided. A method for combining data from a gyroscope and odometry is presented in [19]. The gyroscope improves the angle estimates and thus the final position estimate. The authors of [57] present interesting results for an underground mining vehicle. They clearly show how inertial sensors can be used to correct nonsystematic errors due to soil irregularities when fused with other sensors such as wheel encoders and laser scanners. In [24], the authors propose to use the nonholonomic constraints that govern the motion of a vehicle on a surface, to align the IMU. Finally, 3D-simulation results of a sensor fusion between an IMU and a compass are presented in [55]. The paper states that the system can be used to localize a Mars rover prototype. Unfortunately, the position error grows quickly when localization is done on the sole basis of acceleration and angular velocity integration. Furthermore, a compass cannot be used on Mars because the magnetic field is too weak.

Most of the published works involving IMU on ground vehicles present results in two dimensions and deal with the estimation of the planar position and heading only. Furthermore, the ground is generally flat and the type of soil is known (mostly paved road, carpet or stone). This allows relatively accurate vehicle models to be developed, which yield good odometric information. The conditions in rough terrain are more challenging and these assumptions are not applicable. In particular, no accurate wheel-ground interaction model can be developed and the planar assumption cannot be considered. In this section, the experimental results show how IMU and 3D-Odometry can be combined to provide more reliable motion estimates in three dimensions. As presented in Sect. 5.3, the estimation loop makes use of several models: the state transition model and the measurement models. The methodology to set their covariance matrices is developed here.

Setting the State Transition Covariance Matrix

The parameters of the state transition covariance matrix Q are set using the rover's properties, experimental data and the sensors' datasheets. The parameters of the sub-matrices Q_x, Q_y and Q_z are based on the way the rover is driven and on the general terrain type. We assume that the robot always keeps ground contact, that it is nonholonomic and that the attitude angles are limited to values smaller than 40°. As a consequence, the noise sequences of $\mathbf{x}_x, \mathbf{x}_y$ and $\mathbf{x}_z$ are not independent from each other. Indeed, when the rover is accelerating in the x-y plane, both accelerations along x and y are affected. Furthermore, the z coordinate depends on the x and y coordinates (the robot moves in the 2.5D space). Modeling of this cross-correlation is highly complex because it is a function of nonlinear transformations, which are in turn functions of time. However, in order to avoid excessive complexity and benefit from a simpler filter, we have assumed no cross-correlation between these axes.

Some simple heuristics are applied for estimating how given parameters are related to each other and how they are expected to behave as a function of time. For example, the bias affecting an accelerometer changes more slowly than the acceleration itself. Finally, by increasing slightly the variances of the processes, the filter remains consistent for a larger range of situations. Table 5.1 lists all the parameters together with the heuristics that were used in each case. They

Table 5.1. State transition parameters

$\tau_{\ddot{x}} = \tau_{\ddot{y}} = 0.6 \quad \tau_{\ddot{z}} = 0.1$	The experiments show that the z axis is more subject to vibration when the rover is driving. Thus, it has to be filtered to a greater extent as compared to x and y.
$\sigma^2_{\ddot{x}} = \sigma^2_{\ddot{x}} = 0.008 \quad \sigma^2_{\ddot{z}} = 0.003$	The acceleration along the z axis is generally smaller than the acceleration along x and y axes because the motors of the rover directly affect the acceleration in x and y. The acceleration along z is only due to slope changes in the terrain.
$\tau_{bax} = \tau_{bax} = 0.016 \quad \tau_{baz} = 0.002$	The biases change more slowly than the accelerations, over time. Thus, their time constants are set shorter.
$\sigma^2_{bax} = \sigma^2_{bay} = 0.2 \quad \sigma^2_{baz} = 0.11$	These values are set being equal to the square of half of the maximum biases of the accelerometers (2σ), the values of which are given in the IMU datasheet.
$\tau_{\omega x} = \tau_{\omega y} = 1 \quad \tau_{\omega z} = 3$	ω_z is directly governed by the command signals to the rover. It is thus subject to change more rapidly than $\omega_{x,y}$.
$\sigma^2_{b\omega x} = \sigma^2_{b\omega y} = 0.0006 \; \sigma^2_{b\omega z} = 0.012$	These values are set being equal to the square of half the maximum biases for the gyroscopes (2σ), the values of which can be found in the IMU datasheet.
$\tau_{\Delta\omega z} = 3 \times 10^{-4} \; \sigma^2_{\Delta\omega z} = 3 \times 10^{-5}$	The same reasoning used for setting the acceleration biases is applicable here. According to the IMU datasheet, the scaling factor is less than 1%. So, we took the square of half of this value to set the variance of the scaling factor.
$\tau_{\Delta ox} = \tau_{\Delta oz} = 2 \; \sigma^2_{\Delta ox} = \sigma^2_{\Delta oz} = 1$	These values were determined experimentally. However, the filter is not very sensitive to their variation.

might not be the optimal parameters but they have proven to give good filter performance.

Setting R_{imu}, R_{inc} and R_{odo}

In order to set the uncertainty model of the IMU, the rover was driven forward at different velocities and on different types of soil while collecting data and computing statistics. The experiments showed that the variance of the signals does not change significantly with change in velocity or terrain type. This can be attributed to the passive mechanical structure of SOLERO, which allows for filtering and smoothing of the trajectories. Thus, the worst-case set of variances was selected to set the matrix R_{imu}. For the inclinometer, the variances of the roll and pitch angles were set to the square of half the value given in the IMU datasheet ($2\sigma = 1°$). These values correspond to the diagonal elements of R_{inc}.

The uncertainty model of the odometry is difficult to assess because it depends on the type of terrain. As it is not possible to classify all terrain types, we set the uncertainty of the odometry as being proportional to the acceleration undergone by the rover. Indeed, slip mostly occurs when negotiating an obstacle, while the robot is subject to acceleration. Furthermore, at constant speed, the acceleration is zero and does not bring much information about the rover's motion. In this particular case, motion estimation can only rely on odometry. For the same reasons, the variance for the yaw angle was set proportional to the angular rate ω_z. Thus, the covariance matrix associated with 3D-Odometry is written as

$$C_R = \begin{bmatrix} k_x\left(1+\ddot{x}_R^z - g_x\right) & 0 & 0 & 0 \\ 0 & k_y\left(1+\ddot{y}_R^z - g_y\right) & 0 & 0 \\ 0 & 0 & k_z\left(1+\ddot{z}_R^z - g_z\right) & 0 \\ 0 & 0 & 0 & k_\psi\left(1+\omega_z^z\right) \end{bmatrix} \quad (5.41)$$

where k_x, k_y, k_z and k_ψ are proportional gains and g_x, g_y and g_z are the gravitational components in the rover-fixed frame. The gains are set depending on the detected wheel slip, which is obtained using exteroceptive sensors such as visual odometry. This approach proved to work well and was validated during the experiments. The same equations as presented in Sect. 5.2.2 are applied to propagate the covariance matrix C_R through the coordinate system transformation and to finally obtain R_{odo}.

Experimental Validation

In order to test the position-tracking algorithm, the robot was driven across several experimental setups for a predefined distance. For each of five test runs per setup, the 3D-Odometry trajectory and the filtered trajectory[3] were compared. In this section, we present the results corresponding to the setup depicted in Fig. 5.3, which is the most difficult obstacle the rover had to climb during the experiments.

The experimental setup of Fig. 5.3 is very difficult to negotiate for a wheeled rover because of the sharp edges and the low-friction coefficient. A lot of slip occurred during the step climb and the robot bounced on the ground when the rear bogie wheels descended from the obstacle. The different events occuring during

[3] The filtered trajectory is the trajectory estimated by the EIF filter.

Fig. 5.3. Experimental setup along with the corresponding 3D model used to analyze the results. The maximum height of the obstacle is 135 mm and the final height is 45 mm.

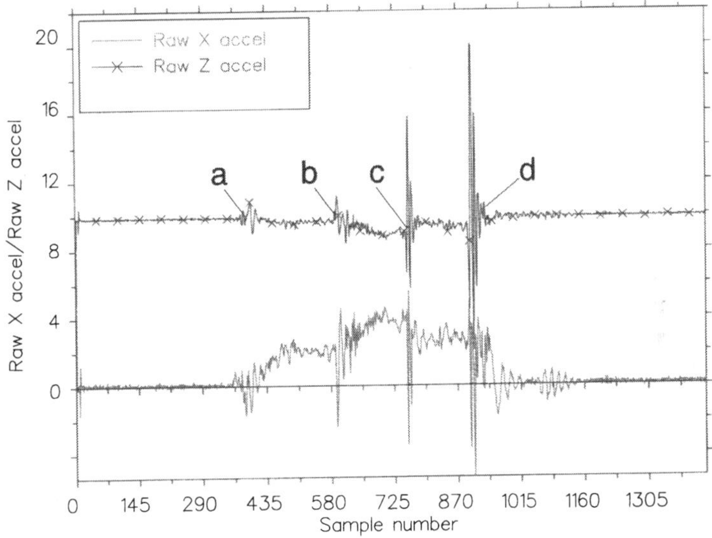

Fig. 5.4. Accelerometers' raw signals for a run performed on the setup depicted in Fig. 5.3: (**a**) front wheels climbing; (**b**) rear wheels climbing; (**c**) front wheels descending; (**d**) rear wheels descending from the concrete blocks

the obstacle negotiation are easily identified when looking at the z-accelerometer signal in Fig. 5.4.

Table 5.4.1 reports the estimation errors of the final position of the trajectories computed with the 3D-Odometry and the filter. The third run, in bold in the table, is used as the reference experiment for the next two figures.

Table 5.2. Experimental results for the setup depicted in Fig. 5.3 (in mm)

Measured			3D-Odometry			Filtered		
x	y	z	x	y	z	x	y	z
1020	4	45	1150	88	40	1160	17	44
1025	7	45	1149	66	40	1152	38	40
1030	**5**	**45**	**1182**	**58**	**38**	**1184**	**18**	**45**
1030	2	45	1149	31	33	1150	29	34
1025	1	45	1152	35	36	1152	16	37
	Mean error		130	52	8	131	20	5

The error along the x-axis is nearly the same for both the filtered and the 3D-Odometry trajectories. This is mainly due to the fact that there are situations where the rover's velocity change is not detected by the accelerometers. Such situations occur in case of wheel slip and when the signal-to-noise ratio of the accelerometers is poor. A typical example is when the rover starts climbing the obstacle. In this case, a lot of wheel slip occurs and the acceleration along x is low. Thus, less weight is given to acceleration and the filter mainly accounts for the odometry measurements. As a consequence, the estimations of the 3D-Odometry and the filtered one do not differ.

On the other hand, when the rover goes down the obstacle the accelerations are significant (Fig. 5.4c and d) and the uncertainty of the odometry increases according to 5.41. Thus, more weight is given to the acceleration measurements and this allows for trajectory correction such as highlighted in Fig. 5.5. For all the experiments the filtered final z-coordinate is always closer to the true height of 45 mm. The error along the z-axis is only 5 mm instead of 8 mm for the 3D-Odometry.

The position error along the y-axis is mainly due to the heading (yaw) error. Indeed, a small heading error translates into large position errors as the traveled distance increases. The error of the heading computed with odometry is mainly due to asymmetric wheel slip, which occurs frequently during this experiment. The heading estimation is significantly improved when the z-gyroscope measurements are integrated. These measurements contain valuable information about the heading rate that is used to filter asymmetric slip and thus to produce better estimates as shown in Fig. 5.6. This results in a significant reduction of the error along the y-axis (see Table 5.4.1).

The position error in the y-z plane is plotted in Fig. 5.7. It clearly appears that the average error of the filtered estimate is closer to the ground-truth (the origin) than the 3D-Odometry estimate. The total error in the plane y-z is 20 mm for the filtered and 52 mm for the 3D-Odometry. Thus, there is an improvement of 44% when using sensor fusion.

For testing the filter in a more general case, the rover was driven several times across the scene depicted in Fig. 5.8. Each time, the operator remote-controlled the rover in order to close the loop. For each run, the final error of the filtered trajectory was smaller than that estimated using odometry only. For the particular

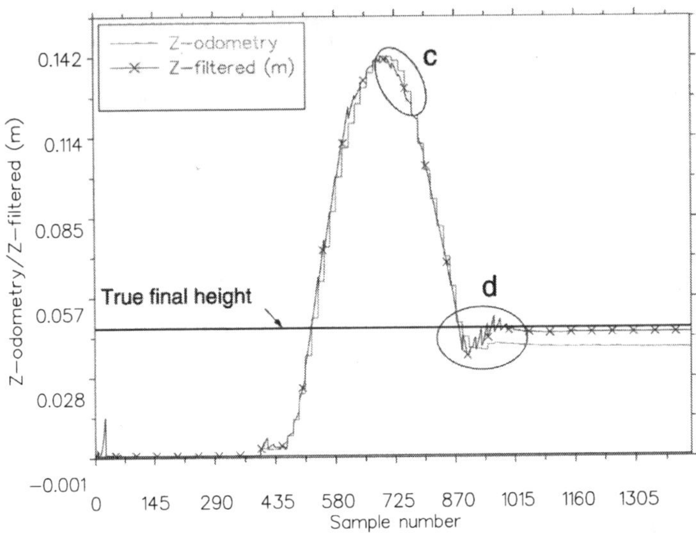

Fig. 5.5. z-coordinate of the trajectory. The acceleration measurements are integrated to correct the trajectory (*circled* areas). The circled areas correspond to events labeled (**c**) and (**d**) in Fig. 5.4.

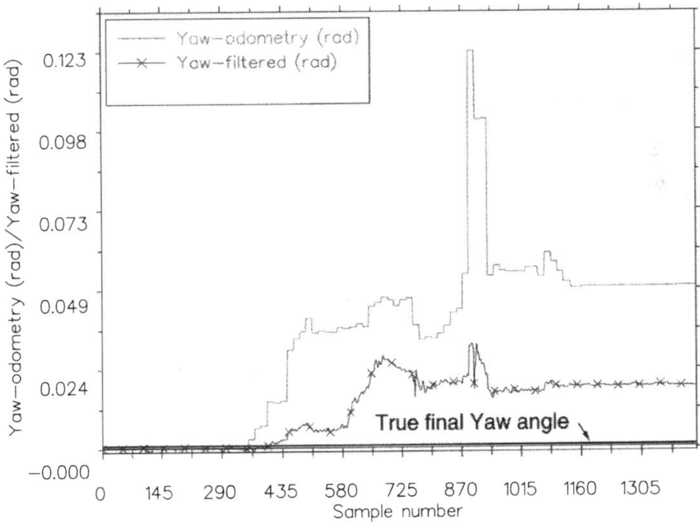

Fig. 5.6. Yaw-angle estimate (the true final angle is close to zero). The yaw gyroscope, measuring the angular rate around z, allows filtering asymmetric slip.

experiment of Fig. 5.8, the final error $[\epsilon_x, \epsilon_y, \epsilon_z, \epsilon_\psi]$ was $[0.16\,m, 0.142\,m, 0.014\,m, 18°]$ for 3D-Odometry and only $[0.06\,m, 0.029\,m, 0.012\,m, 1.2°]$ when using sensor fusion.

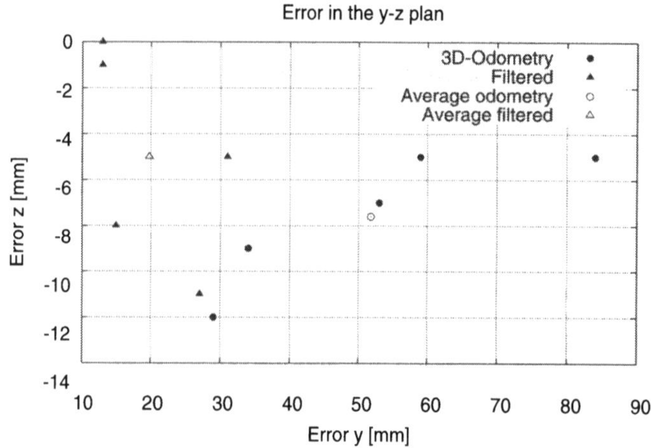

Fig. 5.7. Position error in the y-z plane. The 3D-Odometry estimates are represented by the circles and the filtered estimates by the triangles.

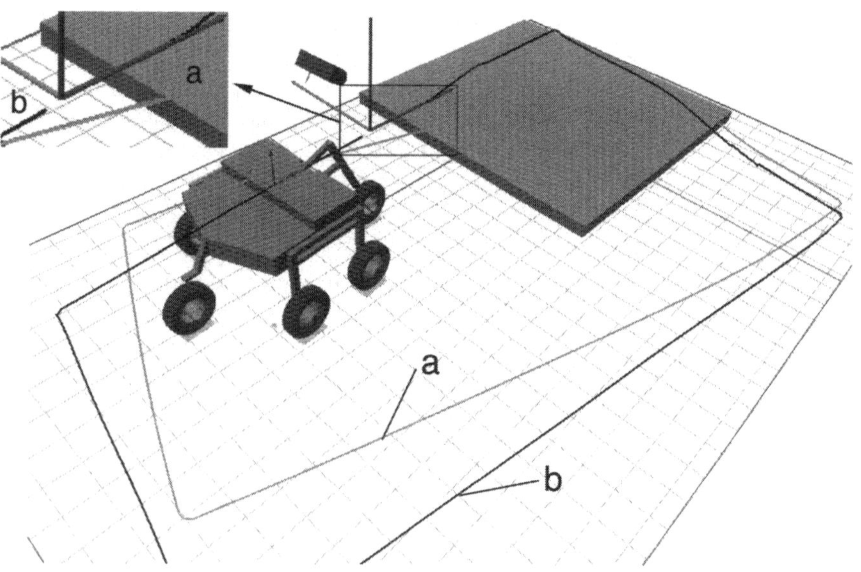

Fig. 5.8. Comparison between (**a**) 3D-Odometry and (**b**) filtered trajectory. This experiment was realized with the real rover.

Discussion

The experimental results show that the inertial measurement unit helps to correct odometric errors and significantly improves the pose estimate. The main contributions occur locally when the robot overcomes sharp-shaped obstacles and during asymmetric wheel slip. In all the experiments, the fusion of odometry and inertial sensors provided better motion estimates than that obtained using odometry only. The improvement brought by the sensor fusion process becomes more and more significant as the total path length increases. In comparison with other works integrating inertial sensors on ground vehicles, this work distinguishes itself with the following features:

- This work addresses the full 3D case.
- An error model for the 3D-Odometry is established based on the IMU measurements according to 5.41.
- The IMU is used in rough terrain, where the signal-to-noise ratio is unfavorable.
- It was shown that the integration of the accelerations can be used to correct the robot's position locally.

5.4.2 Enhancement with VME

In the previous section, only proprioceptive sensors were integrated to track the robot's position. Even if the inertial sensors complete the odometric information, there are situations where the measurements do not provide enough information about the rover's motion. For example, the situation where all the wheels slip, e.g., on ice, is not correctly handled by the filter. In this case, the acceleration is close to zero and the filter only integrates odometric information to produce erroneous position estimates. Thus, in order to increase the robustness of the position tracking and to limit the error growth, it is necessary to incorporate measurements from exteroceptive sensors. These sensors allow computing the rover's ego-motion by tracking features in the environment. This way, wheel slip is detected and correctly handled by the filter.

Here we use the visual motion estimation technique presented in Appendix D to compute the rovers's ego-motion. This new motion information is integrated with the filter using the scheme depicted in Fig. 5.2. The uncertainty model of VME, presented in [36], is based on an error model of stereovision. The covariance matrix R_{vme} expressed in the robot's frame is obtained using the equations in Sect. 5.2.

Experimental Results

One of the experimental setups used to test our position-tracking algorithm is depicted in Fig. 5.9. The use of obstacles of known shape enables the precalculation of reference trajectories (ground-truth trajectories), that are used to assess the accuracy of our approach. The experiment consisted in driving the rover on top of the obstacle while computing the rover's trajectory using the sensor fusion algorithm. A sequence of snapshots taken during the experiment is shown in Fig. 5.10.

The graph of Fig. 5.11 plots four trajectories, i.e. *3D-Odometry*, *VME*, *Estimated* and the *Reference* trajectory. The Estimated trajectory is the result of the sensor fusion of all three sensors. The Reference trajectory was computed considering the kinematics of SOLERO and the geometric dimensions of the experimental setup (the final x coordinate was hand measured). As the rover

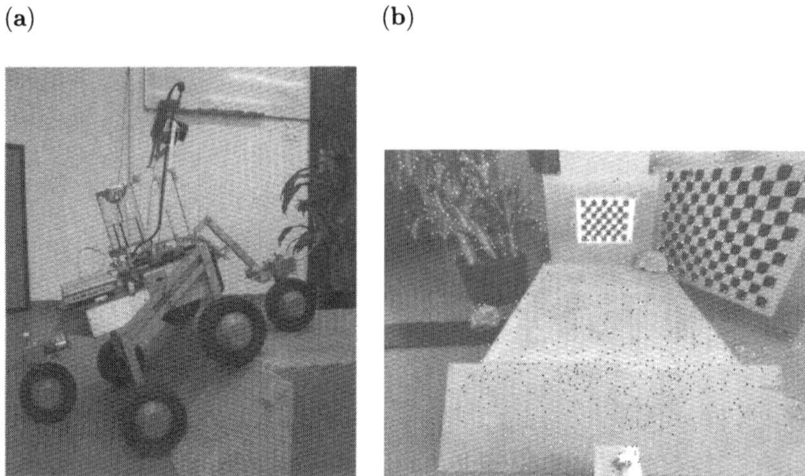

Fig. 5.9. Experimental setup for sensor fusion using VME, 3D-Odometry and IMU: (**a**) side view; (**b**) image taken by the left camera of the stereovision system. The dots represent the extracted Harris features. In comparison with a natural scene, only a few features are detected.

Fig. 5.10. Trajectory of SOLERO (decomposed in three zones)

did not deviate significantly from straight motion, the computed trajectory is considered as the ground-truth.

The graph is divided into three zones, characterizing three different phases of the trajectory as depicted in Fig. 5.10. In zone A, the VME trajectory is very close to the Reference trajectory, whereas the Estimated trajectory is slightly

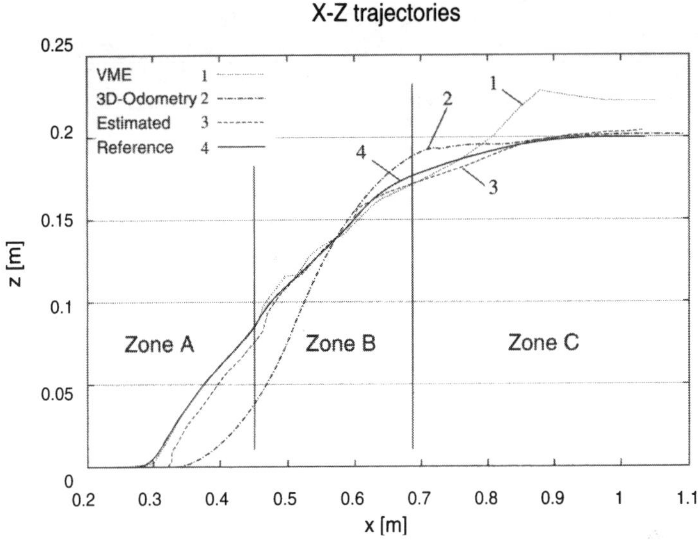

Fig. 5.11. x-z trajectories for the experimental setup of Fig. 5.9

offset. This offset is mainly due to the great amount of wheel slip occurring in zone A. Because the amplitude of the acceleration is low in that part of the trajectory, the weight given to the odometry is not significantly decreased. Thus, the integration of the odometry measurements contributes to push the Estimated trajectory slightly away from the Reference.

In zone B, the Estimated trajectory is closer to the Reference trajectory than VME (see Fig. 5.12). When climbing the obstacle, the motion estimates provided by VME are less accurate because of the lower number of features matched between successive frames. The low number of matches also increases the uncertainty associated with the VME estimates. This explains why the jagged steps of the VME trajectory are not propagated to the level of the Estimated trajectory.

Zone C corresponds to the phase of the trajectory where the rear wheel passes from the sloped surface to the top plane. This transition causes the rover to tilt forward rapidly and thus creates a high discrepancy between two successively acquired images. In such conditions, feature matching is more difficult and only a few pairings are established. The worst case occurs between images number 31 and 32 where only 29 pairings are found. As a consequence, the VME provided a very bad motion estimate with a high uncertainty (see Figs. 5.13 and 5.14). In this case, less weight is given to VME and the sensor fusion can perfectly filter this bad information to produce a reasonably good estimate using the odometry and the IMU instead. Finally, the estimated final position is very close to the measured final position. A final error of four millimeters for a trajectory longer than one meter (0.4%) is very satisfactory, given the difficulty of the terrain (high wheel slip and field-of-view changes).

It is interesting to analyze the plot of the variance of the estimated x coordinate as a function of time. As shown in Fig. 5.15, the variance increases globally,

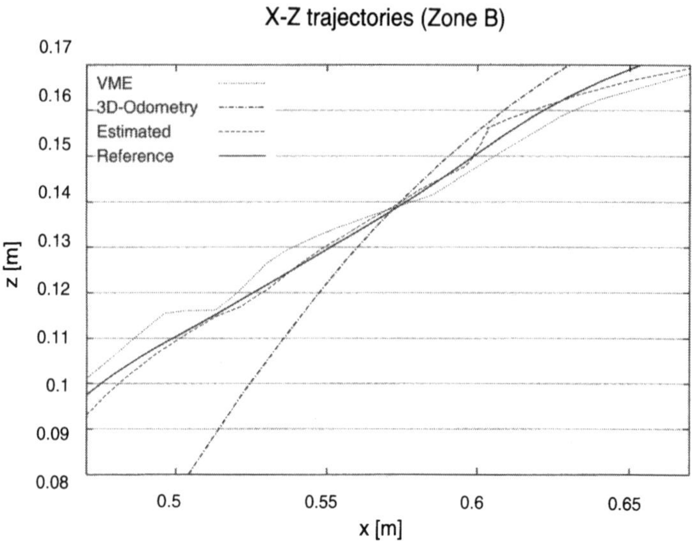

Fig. 5.12. Enlarged view of zone B. The Estimated trajectory is closer to the Reference trajectory than VME.

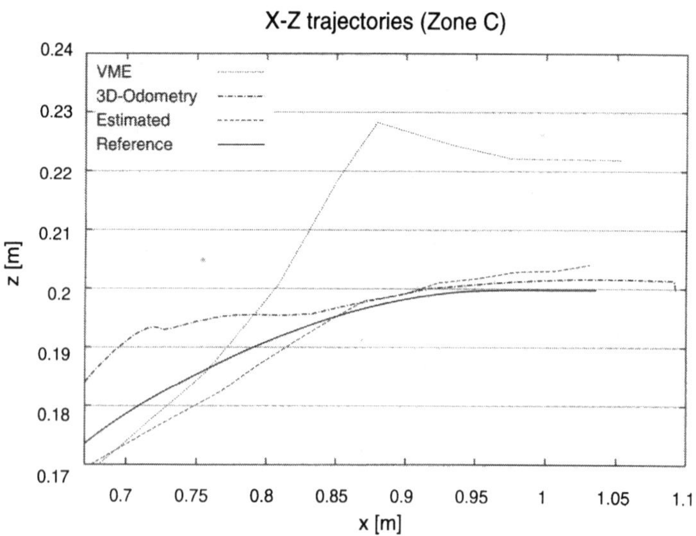

Fig. 5.13. Enlarged view of zone C. Twenty-nine features were matched between images 31 and 32. This yields a bad motion estimate of VME, associated with a large uncertainty. Thanks to sensor fusion, the system successfully filtered this information to provide a good position estimation.

as a function of time. This is because no absolute information is available to reset the position in the global reference frame. The "saw" shape of the plot at the global level is due to the VME updates, whereas the "saw" shape at a local

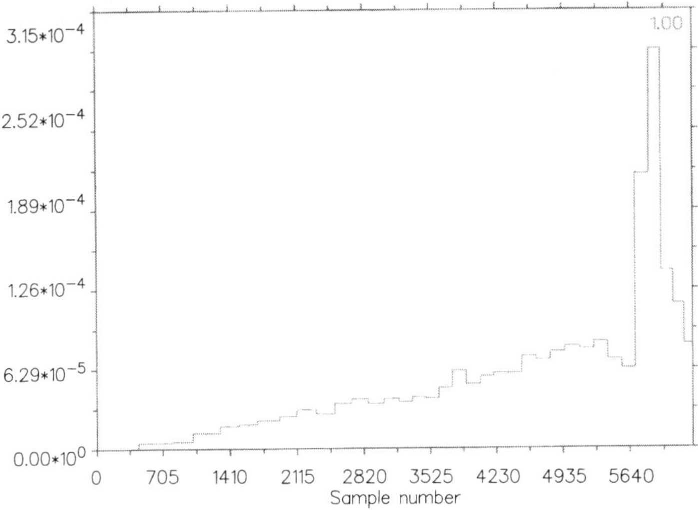

Fig. 5.14. Variance of the motion estimations of VME along x (in meters). The uncertainty increases suddenly at image 32 because only a few features were matched (only 29 pairings).

level is due to the odometry updating the position estimated on the sole base of the IMU measurements. In other words, the motion estimates of VME have the biggest weight, followed, respectively, by the 3D-Odometry and the IMU. In Fig. 5.15, we can also observe the effects of the filter updates when uncertain VME estimations are provided. When the uncertainty of VME is large (see Fig. 5.14), the estimations have less weight and the variance along x remains high. Such filter behaviors are expected and prove that the filter provides consistent estimates.

Note about SLAM

The visual motion estimation technique presented in this chapter enables the computation of the incremental motion between two stereo pairs' acquisitions. Thus, the position error grows each time a new stereo pair is integrated. It is possible to limit the error growth by applying the SLAM (Simultaneous Localization And Mapping) technique [7, 45]. Basically, the approach consists in simultaneously estimating the rover's position while creating a map, composed of relevant features of the environment. By constantly re-observing and matching the same features it is then possible to bound the position error. However, in practice, the method is difficult to apply in rough terrain. Indeed, there is no guarantee of constantly re-observing and matching the same features as the robot moves. The main difficulties are related to occlusions and potentially large field-of-view changes between two data acquisitions. Furthermore, the perception of the environment can be very different when going from position A to B and back from B to A, thus making the problem of feature matching potentially

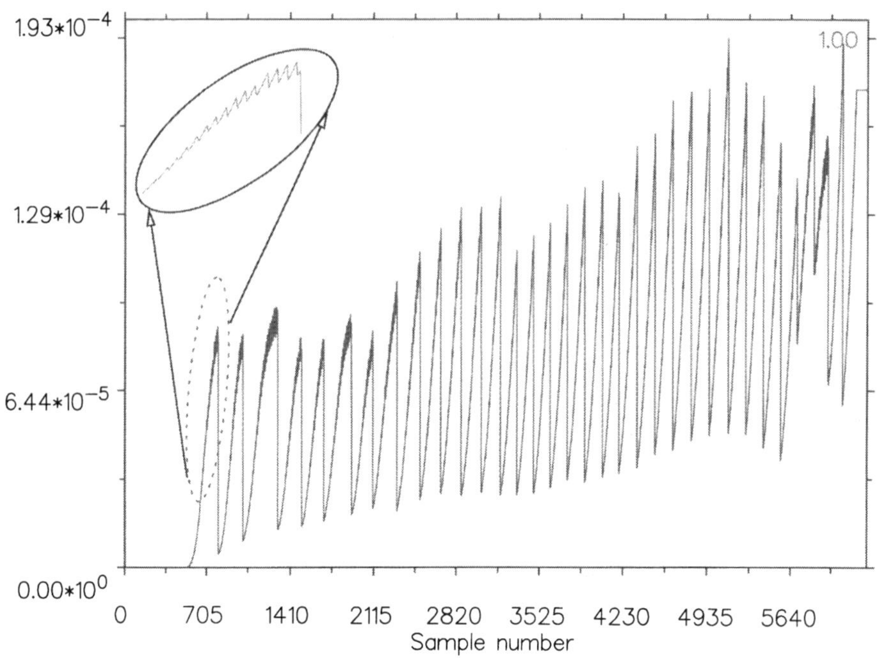

Fig. 5.15. The variance of the position estimates along the x-axis. Because no absolute positioning mechanism is available to reset the position, the variance globally increases over time. The variance significantly decreases each time a VME measurement is available. At a lower level (*circled*), the odometry periodically corrects the position estimates computed using IMU measurements only.

difficult. Another limitation of SLAM is related to the memory requirement when large areas are explored. Finally, in the context of an exploration mission such as the MER mission, the rover does not come back to its initial position, and thus, SLAM does not provide a bounded positioning error anyway.

In spite of all these limitations, SLAM can nevertheless be used locally to refine the motion estimates. Indeed, even if the features are re-observed only a few times and discarded when they disappear from the field-of-view, these multiple observations help to limit the error growth of the position estimate.

5.5 Summary

In this chapter, a robust method for combining different sensor measurements to track the rover's position has been presented. The sensor fusion scheme is flexible and can accommodate any number of sensors of any kind. In order to test the approach, a set of experiments was performed. Three different types of sensors were used, i.e., 3D-Odometry, inertial sensors and a visual motion estimation technique based on stereovision. It turned out that the use of complementary sensors greatly increases the robustness and accuracy of the position estimates. This work distinguishes itself from other similar research projects in the following aspects:

- Sensor fusion is applied in rough terrain and allows tracking the rover's position in three dimensions.
- Sensor fusion is performed with three sensor types (usually only two types of sensors are used).

6 Conclusion

This book is about 3D position tracking and control for all-terrain robots. Its intent is to contribute toward extending the range of possible areas a robot can explore. The work covers a wide range of problems, from system integration aspects up to the development of high-level control algorithms and models. It is shown that control over both the technological and scientific aspects allows reaching better system performance. Improvement of the accuracy of 3D position tracking is obtained by considering rover locomotion in challenging terrains as a holistic problem. Thus, most of the aspects having an influence on pose tracking are addressed, i.e., mechanical design, locomotion control and sensing.

In Chapter 2, the development of a new all-terrain rover has been presented. The platform, called SOLERO, is equipped with two computers, a stereovision module, an omnidirectional vision system, an inertial measurement unit, numerous sensors and actuators and dedicated electronics for power management. The mechanical structure of SOLERO allows smooth motion across rough terrain with limited wheel slip and vibration. This yields good signal-to-noise ratios for the sensors and, in particular, enables the use of inertial sensors in rough terrain. Thus, the intrinsic performance of such a chassis directly contributes to better position tracking.

A method called 3D-Odometry that enables the computation of the rover's motion increments based on wheel encoders and chassis state sensors is presented in Chapter 3. Because it accounts for the rover's kinematic model, this technique provides better motion estimates than the standard approach. The improvement brought by 3D-Odometry becomes particularly significant when considering obstacles comprising sharp edges. Another interesting feature of the approach is that it internally computes the wheel-ground contact angles, which are required inputs for traction control.

To further improve the accuracy of odometry, a predictive controller minimizing wheel slip and maximizing traction has been developed. The approach, presented in Chapter 4, can be applied to various types of terrain because it does not explicitly rely on complex wheel-ground interaction models, whose parameters are generally unknown. Instead of a model-based approach, the controller

P. Lamon: 3D-Position Track. & Cntrl. for All-Terrain Robots, STAR 43, pp. 81–82, 2008.
springerlink.com

uses the sensing capabilities of the rover to estimate rolling resistance as the robot moves. The approach can be adapted to any kind of passive-wheeled rover and runs in real time.

A sensor fusion involving inertial sensors, 3D-Odometry and visual motion estimation is presented in Chapter 5. The experiments demonstrated how each sensor contributes to increase the accuracy and robustness of the 3D pose estimation. In particular, the inertial measurement unit helps to correct odometric errors locally, when the robot overcomes sharp-shaped obstacles and when asymmetric wheel slip occurs. Furthermore, the use of complementary sensors allows increasing the robustness of the tracking algorithm against sensor failure: the rover was able to keep track of its position even in the presence of inaccurate visual motion information.

Because position error grows as a function of the traveled distance, the benefit of combining all these approaches increases as the rover drives longer distances.

A Kinematic and Quasi-static Model of SOLERO

This appendix describes the kinematic model of SOLERO and the quasi-static model that supports the slip minimization algorithm presented in Chapter D.

A.1 Kinematic Model

The chassis of SOLERO is composed of 18 mechanical parts that are numbered as presented in Fig. A.1. The part number 1 is a special part that corresponds to the ground, and thus, is not a mechanical part of SOLERO. The six wheels are numbered as presented in Table A.1.

Table A.1. Wheel numbers and description

Wheel number	Description	Part number
1	Front wheel	13
2	Rear wheel	9
3	Right bogie, front wheel	2
4	Right bogie, rear wheel	3
5	Left bogie, front wheel	14
6	Left bogie, rear wheel	15

In the figures below, the subscript r refers to the robot's frame whereas W refers to the global frame (*world* frame). The label N is used to denote a normal force, T a traction force and T_ω a wheel torque. In order to increase the readability of the figures, the forces between the mechanical parts, i.e., the internal forces, have been omitted. Only the relevant forces and torques are depicted.

P. Lamon: 3D-Position Track. & Cntrl. for All-Terrain Robots, STAR 43, pp. 83–90, 2008.
springerlink.com

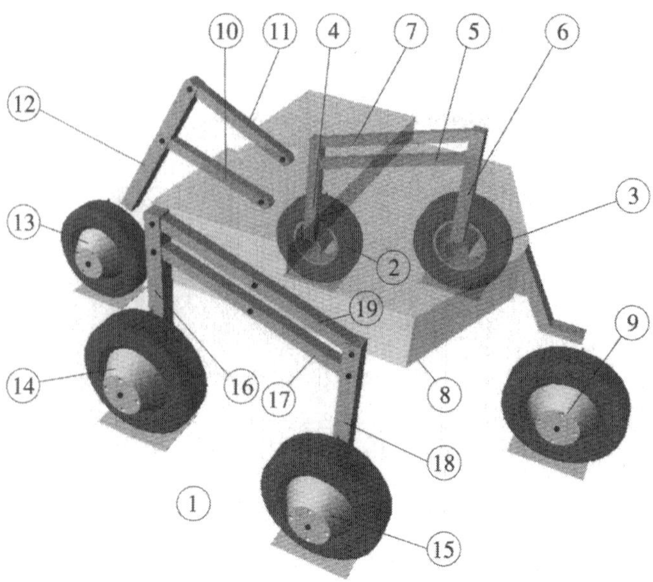

Fig. A.1. Part numbers of SOLERO (part 1 corresponds to the ground)

A.1.1 The Bogies

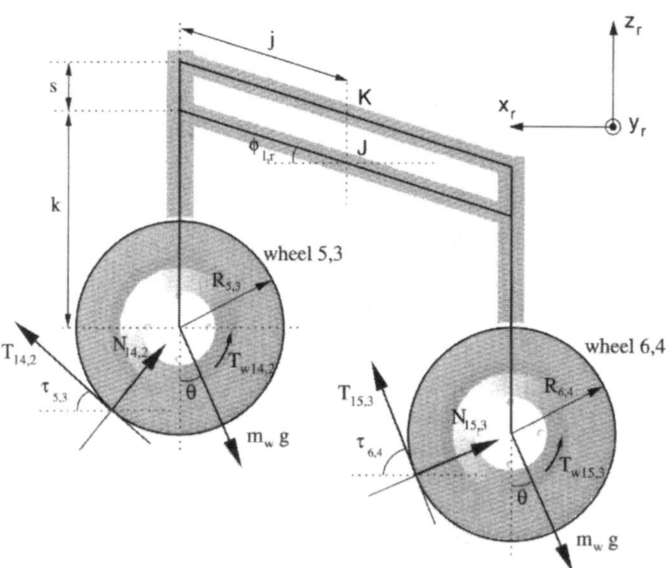

Fig. A.2. Variables and dimensions of a bogie. Subscripts l and r denote the left and right bogie respectively. The bogies are attached to the main body through pin joints placed at points J and K.

A.1.2 The Main Body

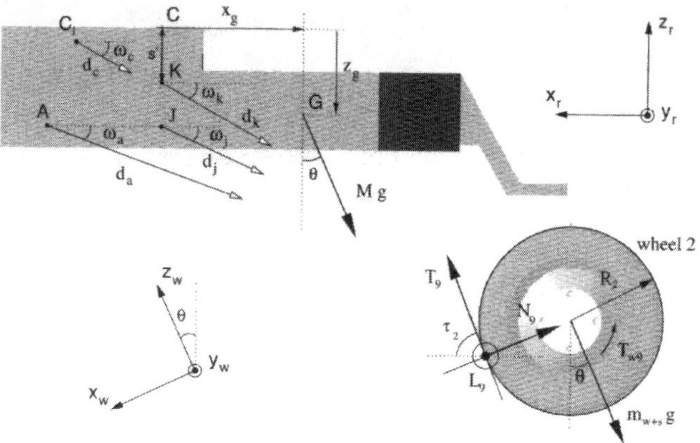

Fig. A.3. Variables and dimensions of the main body

A.1.3 The Front Fork

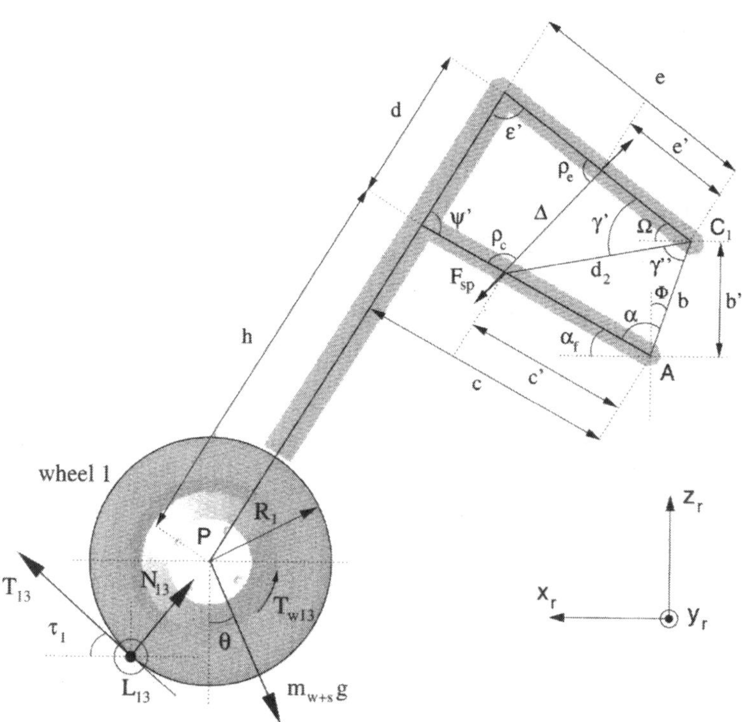

Fig. A.4. Parameters of the front fork. The fork is attached to the main body through points A and C_1.

Trajectory of Point P

The position of the front wheel center P is a function of the angle α_f. Its coordinates are obtained using the parametric equations of angles α and ψ'

$$\alpha(\alpha_f) = \frac{\pi}{2} - \alpha_f + \phi \tag{A.1}$$

$$\begin{aligned}\psi'(\alpha_f) = &\arccos\left(\frac{c - b\cos(\alpha)}{\sqrt{b^2 + c^2 - 2bc\cos(\alpha)}}\right) \\ &+ \arccos\left(\frac{b^2 + c^2 - 2bc\cos(\alpha) + d^2 - e^2}{2d\sqrt{b^2 + c^2 - 2bc\cos(\alpha)}}\right)\end{aligned} \tag{A.2}$$

Finally, the coordinates of P expressed in the robot's frame are given by

$$P(\alpha_f) = \begin{bmatrix} c\cos(\alpha_f) + h\cos(\alpha_f - \psi') \\ c\sin(\alpha_f) - h\sin(\alpha_f - \psi') \end{bmatrix} \tag{A.3}$$

A.2 The Quasi-static Model of SOLERO

The notation used to label the forces and torques follows the standard convention: a generalized force labeled $F_{i,j}$ defines a torque or a force of part i, named F and acting on part j. When $i = 1$ the subscript i is omitted. The following entities are defined to shorten the mathematical expression of the quasi-static model

$$\begin{aligned}
R_x &= \cos(\pi - \alpha_f - \rho_c)\, R \\
R_z &= -\sin(\pi - \alpha_f - \rho_c)\, R \\
mxF &= m_{w+s}\, g\sin(\theta) \\
myF &= -m_{w+s}\, g\sin(\phi)\cos(\theta) \\
mzF &= -m_{w+s}\, g\cos(\phi)\cos(\theta) \\
mxR &= m_{w+s}\, g\sin(\theta) \\
myR &= -m_{w+s}\, g\sin(\phi)\cos(\theta) \\
mzR &= -m_{w+s}\, g\cos(\phi)\cos(\theta) \\
mxBR &= m_w\, g\sin(\theta) \\
myBR &= -m_w\, g\sin(\phi)\cos(\theta) \\
mzBR &= -m_w\, g\cos(\phi)\cos(\theta) \\
mxBL &= m_w\, g\sin(\theta) \\
myBL &= -m_w\, g\sin(\phi)\cos(\theta) \\
mzBL &= -m_w\, g\cos(\phi)\cos(\theta) \\
Mx &= (M - 4\, m_w - 2\, m_{w+s})\, g\sin(\theta) \\
My &= -(M - 4\, m_w - 2\, m_{w+s})\, g\sin(\phi)\cos(\theta) \\
Mz &= -(M - 4\, m_w - 2\, m_{w+s})\, g\cos(\phi)\cos(\theta)
\end{aligned}$$

$$-R_1 T_{13} + Tw_{13} = 0 \qquad -R_2 T_9 + Tw_9 = 0 \qquad -R_3 T_2 + Tw_2 = 0$$
$$-R_4 T_3 + Tw_3 = 0 \qquad -R_5 T_{14} + Tw_{14} = 0 \qquad -R_6 T_{15} + Tw_{15} = 0$$

$$2k \cdot mxBR + F_{7,8x}\, s + Tw_2 + Tw_3 - k\left(N_2 \cos(\tau_3) + N_3 \cos(\tau_4) + T_2 \sin(\tau_3) + T_3 \sin(\tau_4)\right) = 0$$
$$2k \cdot mxBL - F_{8,19x}\, s + Tw_{14} + Tw_{15} - k(N_{14} \cos(\tau_5) + N_{15} \cos(\tau_6) + T_{14} \sin(\tau_5) + T_{15} \sin(\tau_6)) = 0$$
$$T_{8,10x} + h \cdot (L_{13} + myF) \sin(\xi) - c\left((L_{13} + myF) \sin(\alpha_f) + L_{13} \cdot R_1 \cdot \sin(\tau_1)\right) = 0$$
$$T_{8,10z} + c \cdot (L_{13} + myF) \cos(\alpha_f) + L_{13} \cdot R_1 \cdot \cos(\tau_1) + h \cdot (L_{13} + myF) \cos(\xi) = 0$$
$$e \cdot F_{8,11z} \cos(\Omega) + (e' - e) \cdot Fspz \cos(\Omega) - e \cdot F_{8,11x} \sin(\Omega) - (e' - e) \cdot Fspx \sin(\Omega) = 0$$
$$L_{13} + L_9 + My + 2 \cdot myBL + 2 \cdot myBR + myF + myR = 0$$

$$k \cdot \cos(\phi_r) \cdot (-T_2 \cos(\tau_3) + T_3 \cos(\tau_4) + N_2 \sin(\tau_3) - N_3 \sin(\tau_4))$$
$$+ \left(2k \cdot mxBR + F_{7,8x}\, s + Tw_2 + Tw_3 - 2k \cdot (N_2 \cos(\tau_3) + T_2 \sin(\tau_3))\right) \sin(\phi_r) = 0$$

$$k \cos(\phi_l)\,(-T_{14} \cos(\tau_5) + T_{15} \cos(\tau_6) + N_{14} \sin(\tau_5) - N_{15} \sin(\tau_6))$$
$$+ \left(2k \cdot mxBL - F_{8,19x}\, s + Tw_{14} + Tw_{15} - 2k\,(N_{14} \cos(\tau_5) + T_{14} \sin(\tau_5))\right) \sin(\phi_l) = 0$$

$$d \cdot (F_{8,11z} - Fspz) \cos(\xi) + d \cdot F_{8,11x} \sin(\xi) + h \cdot N_{13} \sin(\tau_1 + \xi)$$
$$- \left(Tw_{13} + h \cdot mzF \cos(\xi) + h \cdot T_{13} \cos(\tau_1 + \xi) + d \cdot Fspx \sin(\xi) + h \cdot mxF \sin(\xi)\right) = 0$$

$$(c' \cdot Fspx + c \cdot (F_{8,11x} + mxF)) \sin(\alpha_f)$$
$$- \left((c' \cdot Fspz + c\,(F_{8,11z} - Fspz + mzF)) \cos(\alpha_f) + c\,(T_{13} \cos(\alpha_f - \tau_1) + Fspx \sin(\alpha_f) + N_{13} \sin(\alpha_f - \tau_1))\right) = 0$$

$$- (Mz + 2 \cdot mzBL + 2 \cdot mzBR + mzF + mzR + T_{13} \cos(\tau_1) + T_9 \cos(\tau_2) + T_2 \cos(\tau_3) + T_3 \cos(\tau_4) + T_{14} \cos(\tau_5) + T_{15} \cos(\tau_6))$$
$$+ N_{13} \sin(\tau_1) + N_9 \sin(\tau_2) + N_2 \sin(\tau_3) + N_3 \sin(\tau_4) + N_{14} \sin(\tau_5) + N_{15} \sin(\tau_6) = 0$$

$$2 \cdot B' \cdot mzBR + T_{8,10x} + My \cdot (c_b + z_g) + L_9 \cdot R_2 \sin(\tau_2) + B'(T_2 \cos(\tau_3) + T_3 \cos(\tau_4) + N_{14} \sin(\tau_5) + N_{15} \sin(\tau_6)) - (2k(myBL + myBR)$$
$$+ 2 \cdot B' \cdot mzBL + Mz \cdot y_g + B'(T_{14} \cos(\tau_5) + T_{15} \cos(\tau_6) + N_2 \sin(\tau_3) + N_3 \sin(\tau_4)) + d_j(L_9 + myR) \sin(\omega_j)) = 0$$

$$B'\,((F_{7,8x} + F_{8,19x})s + Tw_2 + Tw_3) + k \cdot (T_{8,10z} + Mx \cdot y_g) + d_j \cdot k(L_9 + myR) \cos(\omega_j)$$
$$- (d_{ja} \cdot k \cdot (L_{13} + myF) + B' \cdot (Tw_{14} + Tw_{15}) + k \cdot My \cdot x_g + k \cdot L_9 \cdot R_2 \cos(\tau_2)) = 0$$

$$- (b' \cdot F_{8,11x} + b'' \cdot F_{8,11z} + d_{ja} \cdot mzF + F_{8,19x}\, s + Tw_9 + Mz \cdot x_g + d_{ja} \cdot T_{13} \cos(\tau_1) + d_j \cdot N_9 \sin(\tau_2 - \omega_j) + d_j \cdot mxR \sin(\omega_j))$$
$$+ F_{7,8x}\, s + Mx \cdot (c_b + z_g) + d_j \cdot T_9 \cos(\tau_2 - \omega_j) + d_j \cdot mzR \cos(\omega_j + d_{ja} \cdot N_{13} \sin(\tau_1)) = 0$$

This set of equations still contains internal forces, i.e., $F_{7,8x}$, $F_{8,10x}$, $F_{8,10z}$, $F_{8,11x}$, $F_{8,11z}$ and $F_{8,19x}$. These unknowns are removed using the Gauss–Jordan elimination.

A.2.1 Linear Dependence of the Wheel Torques

The equation system is simplified using a Gauss–Jordan elimination. The final system contains 15 equations with 20 unknowns and is written as

$$Q_{20\times14}\,\mathbf{F}_{14\times1} = \mathbf{A}_{20\times1} - B_{20\times6}\,\mathbf{M}_{6\times1} \tag{A.4}$$

with $\mathbf{F}$ a vector containing the unknown forces and $\mathbf{M}$ the vector containing the torques. The matrices Q, A and B contain the information about the gravity, the rover's geometry and state. Equation A.4 is rewritten to express the forces as a function of the wheel torques

$$\mathbf{F}_{14\times1} = pinv(Q_{14\times20})\,[\mathbf{A}_{20\times1} - B_{20\times6}\,\mathbf{M}_{6\times1}] \tag{A.5}$$

with $pinv(Q)$, the pseudo-inverse of Q. Now the linear dependence of the torques is proven using the null space property.

Null space definition

C is a linear application. We define the null space of C as the set of vectors whose image as a function of C is the null vector

$$C_{nxm}\; null(C)_{mxl} = 0_{nxl} \tag{A.6}$$

One can write

$$[C\; null(C)]^T = null(C)^T\; C^T = 0_{lxn} \tag{A.7}$$

and then

$$\left[C^T\; null(C^T)\right]^T = null(C^T)^T\; C = 0_{nxl} \tag{A.8}$$

Using the property of A.8 applied to A.4, one can write

$$null(Q^T)^T Q\,\mathbf{F} = 0 = null(Q^T)^T\,[\mathbf{A} - B\,\mathbf{M}] \tag{A.9}$$

Rewriting A.9, we obtain

$$null(Q^T)^T\; B\,\mathbf{M} = null(Q^T)^T\;\mathbf{A} \tag{A.10}$$

$$\mathbf{B}'_{1x6}\;\mathbf{M}_{6x1} = A'_{1x1} \tag{A.11}$$

with

$$\mathbf{B}'_{1x6} = null(Q^T)^T\; B \qquad\qquad A'_{1x1} = null(Q^T)^T\; A$$

Equation A.11 proves that the torques are linearly dependent. Furthermore, this confirms that the solution space is of dimension $m-1$ where m is the number of wheels.

A.2.2 Equal Torque Solution

For SOLERO the solution space is of dimension 5. Among all the possible solutions, the set of torque defined by A.12 is a solution of the system.

$$\mathbf{E}_{6x1} = \frac{A'_{1x1}}{\sum_{i=1}^{6} B'_{1x6}(1,i)} \begin{bmatrix} 1 & 1 & 1 & 1 & 1 & 1 \end{bmatrix}^T \tag{A.12}$$

B Linearized Models

The linearized forms of the nonlinear equations of Chap. 5 are developed in this appendix. In what follows, c and s correspond to the cosine and sine functions. A variable over-lined with a bar denotes an operating point value and the variable h corresponds to the sampling time.

B.1 Accelerometers Model

The first order Taylor expansion of 5.25 is used to obtain the linearized model

$$\begin{bmatrix} \ddot{x} \\ \ddot{y} \\ \ddot{z} \end{bmatrix}_R^z = \bar{\Gamma}_{WR} \begin{bmatrix} \ddot{x} \\ \ddot{y} \\ \ddot{z} \end{bmatrix}_W + \nabla\left[\bar{\Gamma}_{WR}\bar{a}\right] \begin{bmatrix} \phi \\ \theta \\ \psi \end{bmatrix} + \begin{bmatrix} b_{ax} \\ b_{ay} \\ b_{az} \end{bmatrix}_R + \nu_a \tag{B.1}$$

$$\text{with } \bar{a} = \left[\bar{\ddot{x}}\ \bar{\ddot{y}}\ \bar{\ddot{z}} - g\right]^T$$

The Jacobian of 5.28 is written as

$$\nabla\left[\bar{\Gamma}_{WR}\,\bar{a}\right] = \left[\begin{matrix} 0 \\ (\bar{\ddot{z}} - g)c\bar{\theta}c\bar{\phi} + \bar{\ddot{y}}(-c\bar{\psi}s\bar{\phi} + c\bar{\phi}s\bar{\theta}s\bar{\psi}) + \bar{\ddot{x}}(c\bar{\phi}c\bar{\psi}s\bar{\theta} + s\bar{\phi}s\bar{\psi}) \\ -(\bar{\ddot{z}} - g)c\bar{\theta}s\bar{\phi} + \bar{\ddot{x}}(-c\bar{\psi}s\bar{\theta}s\bar{\phi} + c\bar{\phi}s\bar{\psi}) + \bar{\ddot{y}}(-c\bar{\phi}c\bar{\psi} - s\bar{\theta}s\bar{\phi}s\bar{\psi}) \end{matrix}\right.$$

$$\begin{matrix} -(\bar{\ddot{z}} - g)c\bar{\theta} - \bar{\ddot{x}}\,c\bar{\psi}s\bar{\theta} - \bar{\ddot{y}}\,s\bar{\theta}s\bar{\psi} \\ \bar{\ddot{x}}\,c\bar{\theta}c\bar{\psi}s\bar{\phi} - (\bar{\ddot{z}} - g)s\bar{\theta}s\bar{\phi} + \bar{\ddot{y}}\,c\bar{\theta}s\bar{\phi}s\bar{\psi} \\ \bar{\ddot{x}}\,c\bar{\theta}c\bar{\phi}c\bar{\psi} - (\bar{\ddot{z}} - g)c\bar{\phi}s\bar{\theta} + \bar{\ddot{y}}\,c\bar{\theta}c\bar{\phi}s\bar{\psi} \end{matrix}$$

$$\left.\begin{matrix} \bar{\ddot{y}}\,c\bar{\theta}c\bar{\psi} - \bar{\ddot{x}}\,c\bar{\theta}s\bar{\psi} \\ \bar{\ddot{y}}(c\bar{\psi}s\bar{\theta}s\bar{\phi} - c\bar{\phi}s\bar{\psi}) + \bar{\ddot{x}}(-c\bar{\phi}c\bar{\psi} - s\bar{\theta}s\bar{\phi}s\bar{\psi}) \\ \bar{\ddot{x}}(c\bar{\psi}s\bar{\phi} - c\bar{\phi}s\bar{\theta}s\bar{\psi}) + \bar{\ddot{y}}(c\bar{\phi}c\bar{\psi}s\bar{\theta} + s\bar{\phi}s\bar{\psi}) \end{matrix}\right] \tag{B.2}$$

P. Lamon: 3D-Position Track. & Cntrl. for All-Terrain Robots, STAR 43, pp. 91–92, 2008.
springerlink.com

B.2 Gyroscopes State Transition

The linear form of the state transition model 5.40 is written as

$$\begin{bmatrix} \omega_x \\ \phi \\ \omega_y \\ \theta \\ \omega_z \\ \psi \end{bmatrix}_{k+1} = F_\omega \begin{bmatrix} \omega_x \\ \phi \\ \omega_y \\ \theta \\ \omega_z \\ \psi \end{bmatrix}_{k} \tag{B.3}$$

where

$$F_\omega = \begin{bmatrix} e^{-\tau_{\omega_x} h} & 0 & 0 & 0 & 0 & 0 \\ h & 1 + h(\bar{\omega}_y \mathrm{c}\bar{\phi} - \bar{\omega}_z \mathrm{s}\bar{\phi}) \tan\bar{\theta} & h\mathrm{s}\bar{\phi}\tan\bar{\theta} & \frac{h}{\mathrm{c}^2\bar{\theta}}(\bar{\omega}_z \mathrm{c}\bar{\phi} + \bar{\omega}_y \mathrm{s}\bar{\phi}) & h\mathrm{c}\bar{\phi}\tan\bar{\theta} & 0 \\ 0 & 0 & e^{-\tau_{\omega_y} h} & 0 & 0 & 0 \\ 0 & h(-\bar{\omega}_z c\bar{\phi} - \bar{\omega}_y s\bar{\phi}) & h\mathrm{c}\bar{\phi} & 1 & -h\mathrm{s}\bar{\phi} & 0 \\ 0 & 0 & 0 & 0 & e^{-\tau_{\omega_z} h} & 0 \\ 0 & \frac{h}{\mathrm{c}\bar{\theta}}(\bar{\omega}_y \mathrm{c}\bar{\phi} - \bar{\omega}_z \mathrm{s}\bar{\phi}) & h\frac{\mathrm{s}\bar{\phi}}{\mathrm{c}\bar{\theta}} & \frac{h}{\mathrm{c}\bar{\theta}}(\bar{\omega}_z \mathrm{c}\bar{\phi} + \bar{\omega}_y \mathrm{s}\bar{\phi})\tan\bar{\theta} & h\frac{\mathrm{c}\bar{\phi}}{\mathrm{c}\bar{\theta}} & 1 \end{bmatrix} \tag{B.4}$$

C The Gauss–Markov Process

The aim of this appendix is to derive the mathematical expressions of a double-integrated Gauss–Markov process P. A Gauss–Markov process is a stochastic process with zero mean and whose autocorrelation function is written as

$$R_P(t) = \sigma^2 e^{-\tau|t|} \tag{C.1}$$

where $1/\tau$ is the time constant of the process and σ^2 its variance. The power spectral density function of P is given by

$$S_P(j\omega) = \int_{-\infty}^{\infty} R_P(t) \cdot e^{-j\omega t} dt = \frac{2\sigma^2\tau}{\omega^2 + \tau^2} \tag{C.2}$$

A Gauss–Markov process can be considered as a white noise being filtered by a low pass filter with transfer function

$$H(j\omega) = \frac{\sqrt{2\sigma^2\tau}}{j\omega + \tau} \tag{C.3}$$

Equation C.3 is derived from the following relationship

$$S_P(j\omega) = |H(j\omega)|^2 S_U(j\omega) \tag{C.4}$$

where $S_U(j\omega)$ is a unity white noise signal. Figure C.1 depicts the double integration of a Gauss–Markov process p_1. The signal $u(t)$ is a unity white noise and p_2 and p_3 are, respectively, the first and second integrals of p_1. The transfer functions between u and p_1, p_2 and p_3 are, respectively,

$$H_1(s) = \frac{\sqrt{2\sigma^2\tau}}{s+\tau} \qquad H_2(s) = \frac{\sqrt{2\sigma^2\tau}}{s(s+\tau)} \qquad H_3(s) = \frac{\sqrt{2\sigma^2\tau}}{s^2(s+\tau)} \tag{C.5}$$

P. Lamon: 3D-Position Track. & Cntrl. for All-Terrain Robots, STAR 43, pp. 93–95, 2008.
springerlink.com

and the corresponding impulse responses are given by

$$\begin{aligned} h_1(t) &= \sqrt{2\sigma^2\tau}\, e^{-\tau t} \\ h_2(t) &= \sqrt{\frac{2\sigma^2}{\tau}}\,(1 - e^{-\tau t}) \\ h_3(t) &= \sqrt{\frac{2\sigma^2}{\tau^3}}\,(\tau t - 1 + e^{-\tau t}) \end{aligned} \tag{C.6}$$

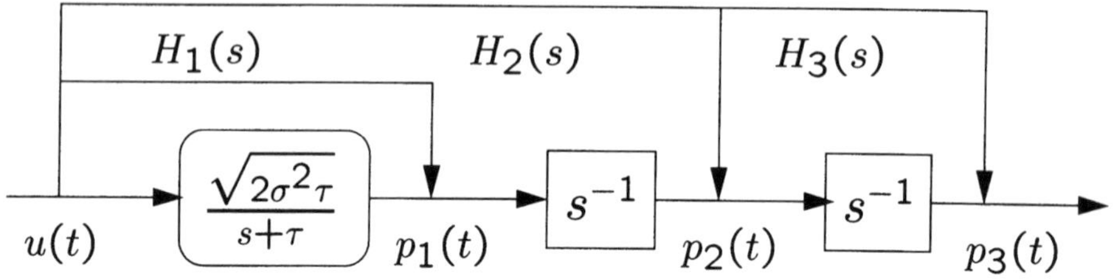

Fig. C.1. Double integration of a Gauss–Markov process

The continuous state transition model of a double-integrated Gauss–Markov process is finally written as

$$\begin{bmatrix} \dot{p}_1 \\ \dot{p}_2 \\ \dot{p}_3 \end{bmatrix} = \begin{bmatrix} -\tau & 0 & 0 \\ 1 & 0 & 0 \\ 0 & 1 & 0 \end{bmatrix} \begin{bmatrix} p_1 \\ p_2 \\ p_3 \end{bmatrix} + \begin{bmatrix} \sqrt{2\sigma^2\tau} \\ 0 \\ 0 \end{bmatrix} u(t) \tag{C.7}$$

The discrete state transition matrix Φ is obtained by applying the inverse Laplace operator on its continuous form

$$\begin{aligned} \Phi = \mathcal{L}^{-1}\left[(sI - F)^{-1}\right] &= \mathcal{L}^{-1} \begin{bmatrix} s+\tau & 0 & 0 \\ -1 & s & 0 \\ 0 & -1 & s \end{bmatrix}^{-1} \\ &= \mathcal{L}^{-1} \begin{bmatrix} 1/s+\tau & 0 & 0 \\ 1/s(s+\tau) & 1/s & 0 \\ 1/s^2(s+\tau) & 1/s^2 & 1/s \end{bmatrix} = \begin{bmatrix} e^{-\tau h} & 0 & 0 \\ (1-e^{-\tau h})/\tau & 1 & 0 \\ (\tau h - 1 + e^{-\tau h})/\tau^2 & h & 1 \end{bmatrix} \end{aligned} \tag{C.8}$$

where h is the sampling time. The covariance matrix of the process, Q, is obtained by computing the expectations of p_1, p_2 and p_3. The covariance of two sequences p_i and p_j is written as

$$E\{p_i\, p_j\} = \int_0^h \int_0^h h_i(\lambda)\, h_j(\varepsilon) E[u(\lambda)u(\varepsilon)]\, d\lambda\, d\varepsilon \tag{C.9}$$

If the signal u is a unity white noise, C.9 can be simplified to

$$E\left\{p_i\, p_j\right\} = \int_0^h \int_0^h h_i(\lambda)\, h_j(\varepsilon)\delta(\lambda - \varepsilon)\, d\lambda\, d\varepsilon = \int_0^h h_i(\lambda)\, h_j(\lambda)\, d\lambda \tag{C.10}$$

All the elements of the covariance matrix Q can be computed using C.10. Only $E\{p2\, p2\}$ and $E\{p2\, p3\}$ are derived here. The other terms are obtained in a similar way

$$\begin{aligned} E\left\{p_2\, p_2\right\} &= \int_0^h h_2(\lambda) h_2(\lambda)\, d\lambda = \int_0^h \left[\sqrt{\frac{2\sigma^2}{\tau}}\left(1 - e^{-\tau\lambda}\right)\right]^2 d\lambda \\ &= \frac{2\sigma^2}{\tau}\left[h - \frac{2}{\tau}\left(1 - e^{-\tau h}\right) + \frac{1}{2\tau}\left(1 - e^{-2\tau h}\right)\right] \end{aligned} \tag{C.11}$$

$$\begin{aligned} E\left\{p_2\, p_3\right\} &= \int_0^h h_2(\lambda) h_3(\lambda)\, d\lambda \\ &= \int_0^h \sqrt{\frac{2\sigma^2}{\tau}}\left(1 - e^{-\tau\lambda}\right) \cdot \sqrt{\frac{2\sigma^2}{\tau^3}}\left(\tau\lambda - 1 + e^{-\tau\lambda}\right) d\lambda \\ &= \frac{2\sigma^2}{\tau^2}\left[\frac{\tau}{2}h^2 - h\left(1 - e^{-\tau h}\right) + \frac{1 - e^{-\tau h}}{\tau} - \frac{1 - e^{-2\tau h}}{2\tau}\right] \end{aligned} \tag{C.12}$$

Finally Q becomes

$$Q = \begin{bmatrix} E\left\{p_1\, p_1\right\} & E\left\{p_1\, p_2\right\} & E\left\{p_1\, p_3\right\} \\ E\left\{p_2\, p_1\right\} & E\left\{p_2\, p_2\right\} & E\left\{p_2\, p_3\right\} \\ E\left\{p_3\, p_1\right\} & E\left\{p_3\, p_2\right\} & E\left\{p_3\, p_3\right\} \end{bmatrix} \tag{C.13}$$

D Visual Motion Estimation

This appendix presents the principle of the visual motion estimation technique used for SOLERO. Such a technique allows computing an estimate of the 3D camera motion between two stereo pairs acquisitions[1]. The six displacement parameters are computed on the basis of a set of 3D point-to-point matches established by tracking the pixels in the image sequence acquired by the robot. The following figure summarizes the approach:

1. At time t, a stereo pair is acquired and the Harris corner detector [29] is applied to extract interest points from both images. Then, the point-matching algorithm presented in [35] is used to find correspondences between the interest points in both images. Finally, the cloud of 3D points is obtained using stereovision and a second outlier rejection cycle is performed [36]. Stereovision is only computed for the interest points in order to reduce the computation time.
2. At time t+1, a new stereo pair is acquired and the Harris points are again extracted from both images. Then, the correspondences between the interest point extracted in the left images (acquired at time t and t+1) are searched using the same technique as presented in 1.
3. The stereovision is used to compute the cloud of 3D points at time t+1.
4. Finally, the six displacement parameters between t and t+1 are computed using the least square minimization technique presented in [28].

The approach is robust and provides accurate motion estimates. A precision of 1% can be attained after a 100-m traverse. More details about the visual motion estimation technique can be found in [36]. In particular, the error model for the six displacement parameters is presented. This uncertainty model is used in Chapter 5 for sensor fusion.

[1] The algorithm used in this project has been developed at the LAAS (Laboratoire d'Analyse et d'Architecture des Systèmes).

P. Lamon: 3D-Position Track. & Cntrl. for All-Terrain Robots, STAR 43, pp. 97–98, 2008.
springerlink.com

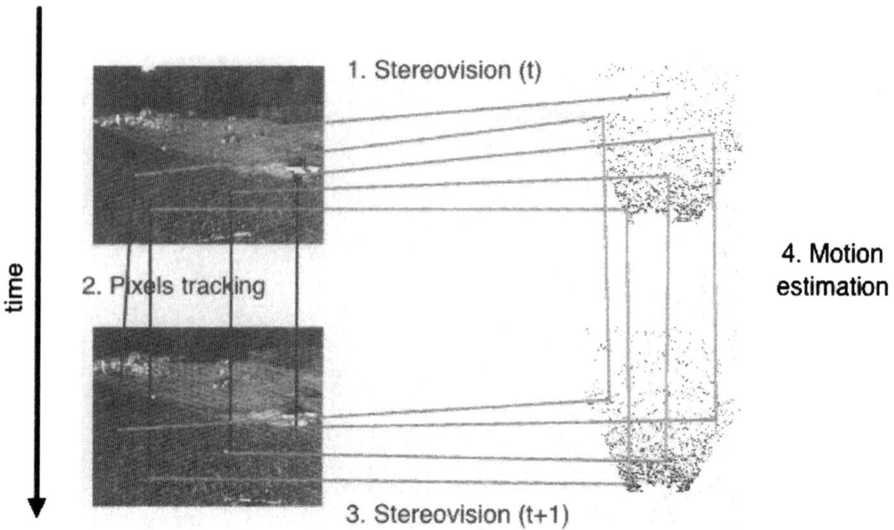

Fig. D.1. Principle of the visual motion estimation (illustration from LAAS)

References

1. Ipc, `http://www.cs.cmu.edu/afs/cs/project/tca/www/ipc/index.html`
2. Life in atacama project, `http://www.frc.ri.cmu.edu/atacama`
3. Ntp, `http://www.ntp.org`
4. Open dynamic engine, `http://www.ode.org`
5. Telemax, `http://www.telerob.de`
6. Andrade, G., Amar, F.B., Bidaud, P., Chatila, R.: Modeling robot-soil interaction for planetary rover motion control. In: IEEE/RSJ International Conference on Intelligent Robots and Systems, Victoria, Canada (1998)
7. Andrade-Cetto, J., Sanfeliu, A.: Environment Learning for Indoor Mobile Robots. In: A Stochastic State Estimation Approach to Simultaneous Localization and Map Building. Springer Tracts in Advanced Robotics, vol. 23, Springer, Heidelberg (2006)
8. Arras, K.O.: An introduction to error propagation: Derivation, meaning and example of equation cy = fx cx ftx. Technical Report EPFL-ASL-TR-98-01 R3, EPFL (2003)
9. Balaram, J.: Kinematic observers for articulated rovers. In: Proceedings of the 2000 IEEE International Conference on Robotics and Automation, San Francisco, CA (2000)
10. Bar-Shalom, Y., Li, X.R.: Estimation and Tracking: Principles, Techniques, and Software. Artech House, Boston, MA (1993)
11. Bares, J., Wettergreen, D.: Lessons from the development and deployment of dante ii. In: Proceedings of the 1997 Field and Service Robotics Conference (1997)
12. Barshan, B., Durrant-Whyte, H.F.: Inertial navigation systems for mobile robots. IEEE Transaction on Robotics and Automation (1995)
13. Baumgartner, E.T., Aghazarian, H., Trebi-Ollennu, A., Huntsberger, T.L., Garrett, M.S.: State estimation and vehicle localization for the fido rover. In: SPIE Proceedings of Sensor Fusion and Decentralized Control in Autonomous Robotic Systems III, Boston, MA, p. 4196 (2000)
14. Bekker, G.: Theory of Land Locomotion, University of Michigan, Ann Arbor (1956)
15. Bekker, G.: Introduction to Terrain-Vehicle Systems. University of Michigan Press (1969)
16. Bertrand, R., Lamon, P., Michaud, S., Schiele, A., Siegwart, R.: The solero rover for regional exploration of planetary surfaces. European Geophysical Society, Geophysical Research Abstracts 5 (2003)

17. Bevly, D.M., Sheridan, R., Gerdes, J.C.: Integrating ins sensors with gps velocity measurements for continuous estimation of vehicle sideslip and tire cornering stiffness. In: Proceedings of the American Control Conference, Arlington (June 2001)
18. Bonnafous, D., Lacroix, S., Simeon, T.: Motion generation for a rover on rough terrains. In: Proceedings of the IEEE/RSJ International Conference on Intelligent Robots and Systems (2001)
19. Borenstein, J., Feng, L.: Measurement and correction of systematic odometry errors in mobile robots. IEEE Journal of Robotics and Automation 12(6), 869–880 (1996)
20. Brooks, C., Iagnemma, K., Dubowsky, S.: Vibration-based terrain analysis for mobile robots. In: IEEE International Conference on Robotics and Automation, Barcelona, Spain (2005)
21. Chahl, J.S., Srinivasan, M.V.: Reflective surfaces for panoramic imaging. Applied Optics 36, 8275–8285 (1997)
22. Cheng, Y., Maimone, M.W., Matthies, L.: Visual odometry on the mars exploration rovers. IEEE Robotics and Automation Magazine (June 2006)
23. Cozman, F., Krotkov, E., Guestrin, C.: Outdoor visual position estimation for planetary rovers. Autonomous Robots 9(2) (2000)
24. Dissanayake, G., Sukkarieh, S., Nebot, E., Durrant-Whyte, H.: The aiding of a low-cost strapdown inertial measurement unit using vehicle model constraints for land vehicle applications. IEEE Transactions on Robotics and Automation 17(5), 731–747 (2001)
25. Estier, T., Crausaz, Y., Merminod, B., Lauria, M., Piguet, R., Siegwart, R.: An innovative space rover with extended climbing abilities. In: Proceedings of Space & Robotics, the Fourth International Conference and Exposition on Robotics in Challenging Environments, Albuquerque, USA (2000)
26. Gancet, J., Lacroix, S.: Pg2p: a perception-guided path planning approach for long range autonomous navigation in unknown natural environments. In: IEEE/RSJ International Conference on Intelligent Robots and Systems, USA (2003)
27. Goldberg, S.B., Maimone, M.W., Matthies, L.H.: Stereo vision and rover navigation software for planetary exploration. In: IEEE Aerospace conference proceedings, Big Sky, Montana, USA, March 2002, vol. 5, pp. 2025–2036 (2002)
28. Haralick, R., Joo, H., Lee, C.N., Zhuang, X., Vaidya, V.G., Kim, M.B.: Pose estimation from corresponding point data. IEEE Transactions on Systems, Man and Cybernetics 19(6), 1426–1445 (1989)
29. Harris, C., Stephens, M.: A combined corner and edge detector. In: Proceedings of the 4th AlveyVision Conference, August 1988, Manchester, pp. 147–151 (1988)
30. Iagnemma, K., Dubowsky, S.: Mobile robot rough-terrain control (rtc) for planetary exploration. In: Proceedings of the ASME Design Engineering Technical Conference, Baltimore, USA (2000)
31. Iagnemma, K., Dubowsky, S.: Vehicle wheelground contact angle estimation: with application to mobile robot traction control. In: 7th International Symposium on Advances in Robot Kinematics, ARK, Piran Portoroz (June 2000)
32. Iagnemma, K., Dubowsky, S.: Mobile Robots in Rough Terrain: Estimation, Motion Planning, and Control with application to Planetary Rovers, June 2004. Springer Tracts in Advanced Robotics (STAR), vol. 12 (2004)
33. Iagnemma, K., Shibley, H., Dubowsky, S.: On-line terrain parameter estimation for planetary rovers. In: IEEE International Conference on Robotics and Automation, Washington DC, USA (2002)

34. Iagnemma, K., Shibly, H., Rzepniewski, A., Dubowsky, S.: Planning and control algorithms for enhanced rough-terrain rover mobility. In: Proceedings of the Sixth International Symposium on Artificial Intelligence, Robotics and Automation in Space, i-SAIRAS, Montreal, Canada (June 2001)
35. Jung, I.-K., Lacroix, S.: A robust interest point matching algorithm. In: International Conference on Computer Vision, Vancouver, Canada (2001)
36. Jung, I.-K., Lacroix, S.: Simultaneous localization and mapping with stereovision. In: International Symposium on Robotics Research, Siena, Italy (2003)
37. Kalker, J.J.: Three dimensional elastic bodies in rolling contact. Kluwer Academic Publishers, Dordrecht (1990)
38. Kemurdjian, A.L., Gromov, V., Mishkinyuk, V., Kucherenko, V., Sologub, P.: Small marsokhod configuration. In: International Conference on Robotics and Automation, Nice (1992)
39. Kubota, T., Kuroda, Y., Kunii, Y., Natakani, I.: Micro planetary rover micro5. In: Proceedings of Fifth International Symposium on Artificial Intelligence,Robotics and Automation in Space (ESA SP-440), Noordwijk, pp. 373–378 (1999)
40. Lacroix, S., Mallet, A., Bonnafous, D., Bauzil, G., Fleury, S., Herrb, M.: Autonomous rover navigation on unknown terrains: functions and integration. International Journal of Robotics Research 21(10–11), 913–942 (2002)
41. Lauria, M.: Nouveaux concepts de locomotion pour véhicules toutterrain robotisés. Thesis nr. 2833, EPFL, Lausanne (2003)
42. Lauria, M., Conti, F., Maeusli, P.-A., Van Winnendael, M., Bertrand, R., Siegwart, R.: Design and control of an innovative micro-rover. In: Proceedings of 5th ESA Workshop on Advanced Space Technologies for Robotics and Automation, Noordwijk (1998)
43. Lauria, M., Piguet, Y., Siegwart, R.: Octopus-an autonomous wheeled climbing robot. In: Proceedings of the Fifth International Conference on Climbing and Walking Robots, Bury St Edmunds and London,UK (2002)
44. Lauria, M., Shooter, S., Siegwart, R.: Topological analysis of robotic n-wheeled ground vehicles. In: Proceedings of the 4th International Conference on Field and Service Robotics, Yamanashi, Japan (2003)
45. Lemaire, T., Berger, C., Jung, I.K., Lacroix, S.: Vision-based slam: Stereo and monocular approaches. International Journal of Computer Vision 74(3) (September 2007)
46. Leppanen, I., Salmi, S., Halme, A.: Workpartner hut automation's new hybrid walking machine. In: CLAWAR 1998 First international symposium, Brussels (1998)
47. Mabie, H.H., Reinholtz, C.F.: Mechanisms and Dynamics of Machinery, 4th edn. John Wiley and Sons, NewYork (1987)
48. Mallet, A., Lacroix, S., Gallo, L.: Position estimation in outdoor environments using pixel tracking and stereovision. In: IEEE International Conference on Robotics and Automation, San Francisco, USA (2000)
49. Manyika, J., Durrant-Whyte, H.: Data fusion and sensor management: A decentralized information-theoretic approach (1994)
50. Michaud, S., Schneider, A., Bertrand, R., Lamon, P., Siegwart, R., van Winnendae, l.M., Schiele, A.: Solero: Solar powered exploration rover. In: Proceedings of the 7 ESA Workshop on Advanced Space Technologies for Robotics and Automation, Noordwijk, The Netherlands (2002)
51. Nebot, E., Sukkarieh, S., Durrant-Whyte, H.: Inertial navigation aided with gps information. In: Proceedings of the Fourth Annual Conference of Mechatronics and Machine Vision in Practice (1997)

52. Ollis, M., Hermann, H., Singh, S.: Analysis and design of panoramic stereo vision using equi-angular pixel cameras. Technical report CMU-RI-TR-99-04, CMU (1999)
53. Olson, C.F., Matthies, L.H., Schoppers, M., Maimone, M.W.: Stereo ego-motion improvements for robust rover navigation. In: IEEE International Conference on Robotics and Automation, Seoul, Corea (2001)
54. Peynot, T., Lacroix, S.: Enhanced locomotion control for a planetary rover. In: Proceedings of the 2003 IEEE/RSJ Intl. Conference on Intelligent Robots and Systems, Las Vegas, USA (2003)
55. Roumeliotis, S.I., Bekey, G.A.: 3d localization for a mars rover prototype. In: 5th International Symposium on Artificial Intelligence, Robotics and Automation in Space (i-SAIRAS 1999), ESTEC, The Netherlands (1999)
56. Roumeliotis, S.I., Johnson, A.E., Montgomery, J.F.: Augmenting inertial navigation with image-based motion estimation. In: IEEE International Conference on Robotics and Automation, Washington, USA (2002)
57. Scheding, S., Dissanayake, G., Nebot, E.M., Durrant-Whyte, H.: An experiment in autonomous navigation of an underground mining vehicle. IEEE Transactions on Robotics and Automation 15(1), 85–95 (1999)
58. Siegwart, R., Estier, T., Crausaz, Y., Merminod, B., Lauria, M., Piguet, R.: Innovative concept for wheeled locomotion in rough terrain. In: Proceedings of the Sixth International Conference on Intelligent Autonomous Systems, Venice, Italy (2000)
59. Siegwart, R., Lamon, P., Estier, T., Lauria, M., Piguet, R.: Innovative design for wheeled locomotion in rough terrain. Journal of Robotics and Autonomous Systems 40(2–3), 151–162 (2002)
60. Singh, S., Simmons, R., Smith, T., Stentz, A., Verma, V., Yahja, A., Schwehr, K.: Recent progress in local and global traversability for planetary rovers. In: IEEE International Conference on Robotics and Automation, Francisco, USA (2000)
61. Stone, H.W.: Mars pathfinder microrover: A low-cost,low-power spacecraft. In: Proceedings of the 1996 AIAA Forum on Advanced Developments in Space Robotics, Madison WI (1996)
62. Strelow, D., Mishler, J., Singh, S., Herman, H.: Extending shape-frommotion to noncentral onmidirectional cameras. In: IEEE/RSJ International Conference on Intelligent Robots and Systems, Hawaii, USA (2001)
63. Strelow, D., Singh, S.: Online motion estimation from image and inertial measurements. In: 11th International Conference on Advanced Robotics, Portugal (2003)
64. Thueer, T., Lamon, P., Krebs, A., Siegwart, R.: Crab - exploration rover with advanced obstacle negociation. In: Proceedings of the 9th ESA workshop on Advanced Space Technologies for Robotics (2006)
65. Titterton, D.H., Weston, J.L.: Strapdown inertial navigation technology. In: Navigation and Avionics, Stevenage (1997)
66. Tomatis, N.: Hybrid, Metric - Topological, Mobile Robot Navigation. Thesis nr. 2444, EPFL (2001)
67. Trebi-Ollennu, A., Huntsberger, T., Yang, C., Baumgartner, E.T., Kennedy, B., Schenker, P.: Design and analysis of a sun sensor for planetary rover absolute heading detection. IEEE Transactions on Robotics and Automation 17(6), 939–947 (December)
68. Tunstel, E.: Evolution of autonomous self-righting behaviors for articulated nanrovers. In: Proceedings of Fifth International Symposium on Artificial Intelligence, Robotics and Automation in Space (ESA SP-440), Noordwijk, pp. 341–346 (1999)

69. van der Burg, J., Blazevic, P.: Anti-lock braking and traction control concept for all-terrain robotic vehicles. In: Proceedings of the 1997 IEEE International Conference on Robotics and Automation, Albuquerque, USA (April 1997)
70. Van Winnendael, M., Visenti, G., Bertrand, R., Rieder, R.: Nanokhod microrover heading towards mars. In: Proceedings of Fifth International Symposium on Artificial Intelligence Robotics and Automation in Space (ESA SP-440), Noordwijk, pp. 69–76 (1999)
71. Vieville, T., Romann, F., Hotz, B., Mathieu, H., Buffa, M., Robert, L., Facao, P.E.D.S., Faugeras, O.D., Audren, J.T.: Autonomous navigation of a mobile robot using inertial and visual cues. In: IEEE/RSJ International Conference on Intelligent Robots and Systems, Tokyo, Japan (1993)
72. Volpe, R., Balaram, J., Ohm, T., Ivlev, R.: Rocky 7: A next generation mars rover prototype. Journal of Advanced Robotics 11(4) (1997)
73. Wada, M., Kang, S.Y., Hashimoto, H.: High accuracy multisensor road vehicle state estimation. In: 26th Annual Conference of the IEEE Industrial Electronics Society, IECON (2000)
74. Yoshida, K., Hamano, H., Watanabe, T.: Slip-based traction control of a planetary rover. In: Proceedings of the 8th International Symposium on Experimental Robotics, ISER, Italy (2002)

Index

Springer Tracts in Advanced Robotics

Edited by B. Siciliano, O. Khatib and F. Groen

Vol. 43: Lamon, P.
3D-Position Tracking and Control for All-Terrain Robots
105 p. 2008 [978-3-540-78286-5]

Vol. 42: Laugier, C.; Siegwart, R. (Eds.)
Field and Service Robotics
300 p. 2008 [978-3-540-75403-9]

Vol. 41: Milford, M.J.
Robot Navigation from Nature
194 p. 2008 [978-3-540-77519-5]

Vol. 40: Birglen, L.; Laliberté, T.; Gosselin, C.
Underactuated Robotic Hands
241 p. 2008 [978-3-540-77458-7]

Vol. 39: Khatib, O.; Kumar, V.; Rus, D. (Eds.)
Experimental Robotics
563 p. 2008 [978-3-540-77456-3]

Vol. 38: Jefferies, M.E.; Yeap, W.-K. (Eds.)
Robotics and Cognitive Approaches to Spatial Mapping
328 p. 2008 [978-3-540-75386-5]

Vol. 37: Ollero, A.; Maza, I. (Eds.)
Multiple Heterogeneous Unmanned Aerial Vehicles
233 p. 2007 [978-3-540-73957-9]

Vol. 36: Buehler, M.; Iagnemma, K.; Singh, S. (Eds.)
The 2005 DARPA Grand Challenge – The Great Robot Race
520 p. 2007 [978-3-540-73428-4]

Vol. 35: Laugier, C.; Chatila, R. (Eds.)
Autonomous Navigation in Dynamic Environments
169 p. 2007 [978-3-540-73421-5]

Vol. 34: Wisse, M.; van der Linde, R.Q.
Delft Pneumatic Bipeds
136 p. 2007 [978-3-540-72807-8]

Vol. 33: Kong, X.; Gosselin, C.
Type Synthesis of Parallel Mechanisms
272 p. 2007 [978-3-540-71989-2]

Vol. 32: Milutinović, D.; Lima, P.
Cells and Robots – Modeling and Control of Large-Size Agent Populations
130 p. 2007 [978-3-540-71981-6]

Vol. 31: Ferre, M.; Buss, M.; Aracil, R.; Melchiorri, C.; Balaguer C. (Eds.)
Advances in Telerobotics
500 p. 2007 [978-3-540-71363-0]

Vol. 30: Brugali, D. (Ed.)
Software Engineering for Experimental Robotics
490 p. 2007 [978-3-540-68949-2]

Vol. 29: Secchi, C.; Stramigioli, S.; Fantuzzi, C.
Control of Interactive Robotic Interfaces – A Port-Hamiltonian Approach
225 p. 2007 [978-3-540-49712-7]

Vol. 28: Thrun, S.; Brooks, R.; Durrant-Whyte, H. (Eds.)
Robotics Research – Results of the 12th International Symposium ISRR
602 p. 2007 [978-3-540-48110-2]

Vol. 27: Montemerlo, M.; Thrun, S.
FastSLAM – A Scalable Method for the Simultaneous Localization and Mapping Problem in Robotics
120 p. 2007 [978-3-540-46399-3]

Vol. 26: Taylor, G.; Kleeman, L.
Visual Perception and Robotic Manipulation – 3D Object Recognition, Tracking and Hand-Eye Coordination
218 p. 2007 [978-3-540-33454-5]

Vol. 25: Corke, P.; Sukkarieh, S. (Eds.)
Field and Service Robotics – Results of the 5th International Conference
580 p. 2006 [978-3-540-33452-1]

Vol. 24: Yuta, S.; Asama, H.; Thrun, S.; Prassler, E.; Tsubouchi, T. (Eds.)
Field and Service Robotics – Recent Advances in Research and Applications
550 p. 2006 [978-3-540-32801-8]

Vol. 23: Andrade-Cetto, J,; Sanfeliu, A.
Environment Learning for Indoor Mobile Robots – A Stochastic State Estimation Approach to Simultaneous Localization and Map Building
130 p. 2006 [978-3-540-32795-0]

Vol. 22: Christensen, H.I. (Ed.)
European Robotics Symposium 2006
209 p. 2006 [978-3-540-32688-5]

Vol. 21: Ang Jr., H.; Khatib, O. (Eds.)
Experimental Robotics IX – The 9th International Symposium on Experimental Robotics
618 p. 2006 [978-3-540-28816-9]

Vol. 20: Xu, Y.; Ou, Y.
Control of Single Wheel Robots
188 p. 2005 [978-3-540-28184-9]

Vol. 19: Lefebvre, T.; Bruyninckx, H.; De Schutter, J. Nonlinear Kalman Filtering for Force-Controlled Robot Tasks
280 p. 2005 [978-3-540-28023-1]

Vol. 18: Barbagli, F.; Prattichizzo, D.; Salisbury, K. (Eds.)
Multi-point Interaction with Real and Virtual Objects
281 p. 2005 [978-3-540-26036-3]

Vol. 17: Erdmann, M.; Hsu, D.; Overmars, M.; van der Stappen, F.A (Eds.)
Algorithmic Foundations of Robotics VI
472 p. 2005 [978-3-540-25728-8]

Vol. 16: Cuesta, F.; Ollero, A.
Intelligent Mobile Robot Navigation
224 p. 2005 [978-3-540-23956-7]

Vol. 15: Dario, P.; Chatila R. (Eds.)
Robotics Research – The Eleventh International Symposium
595 p. 2005 [978-3-540-23214-8]

Vol. 14: Prassler, E.; Lawitzky, G.; Stopp, A.; Grunwald, G.; Hägele, M.; Dillmann, R.; Iossifidis. I. (Eds.)
Advances in Human-Robot Interaction
414 p. 2005 [978-3-540-23211-7]

Vol. 13: Chung, W.
Nonholonomic Manipulators
115 p. 2004 [978-3-540-22108-1]

Vol. 12: Iagnemma K.; Dubowsky, S.
Mobile Robots in Rough Terrain – Estimation, Motion Planning, and Control with Application to Planetary Rovers
123 p. 2004 [978-3-540-21968-2]

Vol. 11: Kim, J.-H.; Kim, D.-H.; Kim, Y.-J.; Seow, K.-T.
Soccer Robotics
353 p. 2004 [978-3-540-21859-3]

Vol. 10: Siciliano, B.; De Luca, A.; Melchiorri, C.; Casalino, G. (Eds.)
Advances in Control of Articulated and Mobile Robots
259 p. 2004 [978-3-540-20783-2]

Vol. 9: Yamane, K.
Simulating and Generating Motions of Human Figures
176 p. 2004 [978-3-540-20317-9]

Vol. 8: Baeten, J.; De Schutter, J.
Integrated Visual Servoing and Force Control – The Task Frame Approach
198 p. 2004 [978-3-540-40475-0]

Vol. 7: Boissonnat, J.-D.; Burdick, J.; Goldberg, K.; Hutchinson, S. (Eds.)
Algorithmic Foundations of Robotics V
577 p. 2004 [978-3-540-40476-7]

Vol. 6: Jarvis, R.A.; Zelinsky, A. (Eds.)
Robotics Research – The Tenth International Symposium
580 p. 2003 [978-3-540-00550-6]

Vol. 5: Siciliano, B.; Dario, P. (Eds.)
Experimental Robotics VIII – Proceedings of the 8th International Symposium ISER02
685 p. 2003 [978-3-540-00305-2]

Vol. 4: Bicchi, A.; Christensen, H.I.; Prattichizzo, D. (Eds.)
Control Problems in Robotics
296 p. 2003 [978-3-540-00251-2]

Vol. 3: Natale, C.
Interaction Control of Robot Manipulators – Six-degrees-of-freedom Tasks
120 p. 2003 [978-3-540-00159-1]

Vol. 2: Antonelli, G.
Underwater Robots – Motion and Force Control of Vehicle-Manipulator Systems
268 p. 2006 [978-3-540-31752-4]

Vol. 1: Caccavale, F.; Villani, L. (Eds.)
Fault Diagnosis and Fault Tolerance for Mechatronic Systems – Recent Advances
191 p. 2003 [978-3-540-44159-5]